Planetary Landscapes

Planetary Landscapes

SECOND EDITION

Ronald Greely

CHAPMAN
& HALL

Second edition
Published by
Chapman & Hall
One Penn Plaza
New York, NY 10119

Published in Great Britain by
Chapman & Hall
2-6 Boundary Row
London SE1 84N

Printed in the United States of America

Library of Congress Cataloging-in-Publication Data

Greeley, Ronald.
    Planetary landscapes / Ronald Greeley. — 2nd ed.
        p.   cm.
    Includes index.
        1. Planets—Surfaces.   2. Solar system—Exploration.   I. Title.
    QB603.S95G74   1993
    559.9′2—dc20   93-6885
    ISBN 0-412-05431-0 (hb); ISBN 0-412-05181-8 (pb)   CIP

Please send your order for this or any Chapman & Hall book to **Chapman & Hall, 29 West 35th Street, New York, NY 10001, Attn: Customer Service Department.** You may also call our Order Department at 1-212-244-3336 or fax your purchase order to 1-800-248-4724.

For a complete listing of Chapman & Hall's titles, send your requests to **Chapman & Hall, Dept. BC, One Penn Plaza, New York, NY 10119.**

To Don Gault and Steve Dwornik,
who introduced me to the Moon and planets

Contents

List of tables

Preface

The objective of this book is to introduce the surface features of the planets and satellites in the context of geomorphic processes. Introductory chapters include the "hows" and "whys" of Solar System exploration and a review of the primary processes that shape our planet, Earth, and which appear to be important to planetary sciences. The remaining chapters describe the geomorphology of the planets and satellites for which data are available. For most of these objects, the general physiography and terrain units for each are introduced, then the geomorphic processes that are inferred for the development of their surfaces are described. Each chapter then ends with a synopsis of the geologic evolution of the surface. The principal sources of information on the geomorphology of the planets and satellites are spacecraft photographs. These are usually shown in the book oriented so that the illumination is from the upper left or left side (so that craters will appear correctly as holes to most readers!), although this means that north orientation may not be toward the top.

Because the level of knowledge is not uniform for all of the objects in the Solar System, the individual treatment varies among the chapters. For example, it was difficult to decide what to leave out of the chapter on Mars because so much is known about the surface, whereas data are rather limited for Mercury.

In addition to introducing the geomorphology of planetary objects, this book is intended to be a "source" for obtaining supplemental information. References are cited throughout the text. However, these citations are not intended to be exhaustive but rather are given to provide a "springboard" for additional literature surveys.

Finally, it must be pointed out that planetary sciences are in their infancy and the techniques for analyzing the geomorphology of extraterrestrial objects are still evolving. I hope that the reader will find these extraterrestrial worlds as fascinating and as exciting as I do and, together with my colleagues, we are pleased to share the results of Solar System exploration with our readers.

R. Greeley
September 1992

Acknowledgments

Preparing a book is a formidable project. Without the help of many people it is doubtful that this project would have been completed. I wish to acknowledge the following individuals and thank them for their contributions: Jeffrey Moore and Curtis Manley for research assistance in preparation of the first edition; Loretta Moore for helpful comments on early drafts of the manuscript; Daniel Ball, Joo-Keong Lim, and Joseph Riggio for photographic support; Maureen Geringer for her accurate and speedy word processing of countless draft versions, and Shana Blixt for word processing material for the second edition.

Numerous colleagues very kindly provided photographs and illustrations as noted in the figure captions. Jurie Van der Woude of the Jet Propulsion Laboratory is gratefully acknowledged for providing access to planetary images. The Regional Planetary Image Facilities also aided greatly in providing images for this book and I would especially like to thank Linda Jaramillo and Conni Alexander (Arizona State University), Leslie Pieri and Marion Rudnyk (Jet Propulsion Laboratory), Gail Georgenson (University of Arizona), and Jody Swann (US Geological Survey).

Reviews of individual chapters and helpful discussions were provided by Paul Spudis (US Geological Survey), John Guest (University of London Observatory), Alan Peterfreund and James Garvin (Brown University), George McGill (University of Massachusetts), Peter Schultz (Lunar Planetary Science Institute), Michael Malin (Malin Space Science Systems), Victor Baker (University of Arizona), and Peter Cattermole (University of Sheffield).

I thank Conni Alexander and Andrea Seelinger for their careful proofing of the second edition. Finally, I acknowledge with gratitude the editorial assistance of Cynthia Greeley.

Planetary Landscapes

1 Introduction

The mid-1960s witnessed two fundamental revelations that resulted in the beginning of an era which continues to have profound effects on the geologic sciences. Although the basic ideas of continental drift had been proposed nearly a half century earlier by Alfred Wegener, it was not until the late 1950s and early 1960s and the development of modern instruments to measure sea-floor spreading, to date rocks radiometrically, and to conduct accurate geophysical surveys, that the concept of crustal plate tectonics was accepted. At the same time as the gradual acceptance of plate tectonics, a similar, equally profound, view of the Solar System was emerging. Just as the ideas of continental drift were hampered by the lack of data, Solar System studies were also limited until the space age. Prior to the return of results from space probes sent to the Moon and planets, views of most planetary objects (except the Moon) were limited to little more than fuzzy blurs or tiny pin-points of light, even when viewed with the most powerful Earth-bound telescopes (Fig. 1.1).

Primitive by today's standards, the first probes sent to the Moon by the Soviets ushered in a new scientific era which has brought the surfaces of the planets within the grasp of study. For the most part, the study of planetary surfaces has passed from the astronomer to the geologist and has resulted in the establishment of the new discipline, **planetary geology**. Planetary geology is defined as the study of the origin, evolution and distribution of matter which forms the planets, natural satellites, comets and asteroids. The term "geology" is used in the broadest sense and is considered to mean the study of the solid parts of the planets. Aspects of geophysics, geochemistry, geodesy and cartography are all included in the general term.

This book is concerned with one aspect of planetary geology—the geomorphology of the planets. For simplicity, the term "planet" is also applied to natural satellites, such as the Moon. Geomorphology, a discipline long established in the earth sciences, seeks to determine the form and evolution of the Earth's surface. This same goal

Figure 1.1 Progressively higher-resolution views of Mars.
(a) Earth-based telescopic photograph of the western hemisphere of Mars taken 16 August 1971 by the University of Arizona, Lunar and Planetary Laboratory using the Catalina Observatory Telescope (courtesy of the Lunar and Planetary Laboratory).

(b) A global view of Mars similar to Figure 1.1a, imaged by Mariner 7 on 4 August 1969 from a distance of 495 086 km; outline shows location of Figure 1.1c and d (Mariner 7 7F71).

1

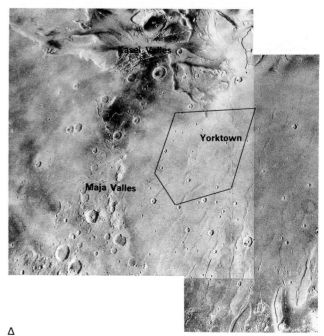

Δ

(d) Viking orbiter images of about same region as shown in Figure 1.1c; the significant improvement in overall picture quality results primarily from an improved imaging system. Area shown is about 1200 km across. The outline locates Figure 1.1e (Viking Orbiter 666A04, 666A06).

(c) Mariner 9 images of the Kasei region of Mars taken from 2835 km; resolution ~1.5 km/pixel; area shown is 650 km by 1000 km; outline shows location of Figure 1.1e (Mariner 9 DAS 07543588 and DAS 07543518).

(e) High-resolution Viking orbiter mosaic of the Kasei region. Rectangle locates Figure 1.1f. Area shown is ~400 km across (Viking Orbiter mosaic 211-5015).

2

(f) High-resolution Viking orbiter image of the Viking 1 landing site taken from 1551 km; resolution ~350 m/pixel. The polygon outline locates Figure 1.1g. Area shown is 46 km across (Viking Orbiter 27A33).

(g) Highest-resolution (10 m/pixel) orbiter images of Viking 1 landing site. Craters near the site can be seen in the lander panorama (Fig. 1.1h); also seen is the crater formed by the impact of the Viking thermal shield (O. de Coursac and J. Garvin, personal communication, 1990). Area shown is about 12 km across (Viking Orbiters 452B11 and 452B10).

(h) Viking lander panorama; rims of craters seen in Figure 1.1g are visible on the horizon (Viking Lander 11A156/027).

3

applies to all the planets. Because much of our knowledge of the planets has been gained by remote sensing—primarily spacecraft pictures—we are, in fact, using landforms to interpret much of the geology of the planets. The approach commonly employed is first to determine the physiography of each planet and then to interpret the processes that have shaped the landforms that are observed on the pictures.

1.1 Objectives of Solar System exploration

The question is often asked, "Why study the Solar System?" During the mid-1960s, the United States National Academy of Sciences addressed this question and defined three principal goals for the exploration of space: (a) to determine the origin and evolution of the Solar System; (b) to determine the origin and evolution of life; and (c) to clarify the nature of the processes that shape humankind's terrestrial environment. The US National Aeronautics and Space Administration (NASA) has refined these objectives and derived a plan for Solar System exploration through the year 2000 (SSEC 1983).

Planetary geology and geomorphology figure prominently in all three of these goals, as discussed in detail by Greeley and Carr (1976). Let us consider the first goal. There are two ways to approach the study of the origin and evolution of the Solar System. The first is to model the possible conditions of Solar System formation and then follow the evolution through a series of stages leading up to the present time. This approach is typically employed by astronomers and uses observations of stars that are thought to represent various stages of evolution. The second approach is to begin with the present state of the Solar System and try to work backward in time. This is the approach typically taken by the planetary geologists—it is basically a geologic approach. Of course, the final goal will probably be reached through a combination of these two approaches. Astronomical modeling is probably the better approach for the early stages of formation and for that part of the early record missing in geologically derived histories; the geologic approach is better for the later stages of Solar System evolution represented by the rock record preserved on the surfaces of planetary objects.

The geology of planetary surfaces also relates to the second goal of Solar System studies, the origin and evolution of life. The planetary environment, including rock and mineral compositions and active geologic processes, has a direct bearing on the starting conditions for life and influences the process of natural selection.

And what of the third goal? Many fundamental geologic problems on Earth might be solved by detailed comparisons with other planets where the relative effects of different sizes, compositions and atmospheres on the evolution of the planets could be assessed. For example, very little is known of the early history of the Earth. Only the last 0.5 eons of the estimated 4.6 eon history is readily available for study because so much of our planet is constantly attacked and altered by tectonic, weathering and erosional processes, and the remainder is covered by water. On the other hand, because the Moon has no erosive atmosphere and the crust has been stable for several eons, it displays a surface that is commonly five to eight times older than most of the Earth's surface. In this older surface is stored and available for study the early history of the Moon and probably the Earth as well. Thus, through the study of the Moon, we are better able to understand the processes that contributed to the early history of the Earth (Fig. 1.2).

To achieve the goals of Solar System exploration, data are obtained through a series of steps that begins with Earth-based observations, progresses through reconnaissance missions and ultimately ends with manned exploration. Obviously, we are a long way from completing this sequence, even for a small fraction of the Solar System, as indicated in Table 1.1. Nevertheless, sufficient data are available to begin analyses and to formulate ideas on the origin and evolution of the planets and their geologic histories.

Table 1.1 Sequence of planetary exploration.

Type of mission	Mercury	Venus	Moon	Mars	Jupiter	Saturn	Uranus	Neptune	Pluto	Comets	Asteroids
Earth-based	x	x	x	x	x	x	x	x	x	x	x
Flyby	x	x	x	x	x	x	x	x		x	x
Orbiter			x	x	x	(1)					
Unmanned landers		x	x	x							
Sample return			x								
Manned landings			x								

(1) Anticipated Galileo mission.

Figure 1.2 Apollo 17 view of the eastern part of the Moon ("farside" is to the right) showing heavily cratered uplands and dark mare areas (circular zone in upper half is Mare Crisium). The samples from the uplands have been dated by radiometric techniques as having been formed in excess of 4 eons ago and thus represent geologic events from the earliest history of the Solar System. Because of the proximity of the Moon, it is presumed that Earth experienced a similar period of heavy cratering, but because of tectonic and erosional processes, most of the early record on Earth has been lost (Apollo 17 AS 17-152-23308).

1.2 The geologic approach

With each step in the exploration of the Solar System, three primary geologic questions are asked: (a) What is the present state of the planetary object observed? (b) What is its geologic history? (c) How do the present state and geologic history compare with other objects in the Solar System? The keys to addressing these questions involve the interpretation of planetary processes, the derivation of geologic histories through mapping and comparative planetology.

1.2.1 Present state

An understanding of the present state of a planetary object requires knowledge of its composition, interior properties, exterior environment and the geologic processes which may be currently active. Knowledge of the composition—at least of the surface—can be obtained directly by measurements made on returned samples (Fig. 1.3), or *in situ* by various instruments on landed spacecraft (Fig. 1.4), or indirectly by various remote-sensing techniques (Fig. 1.5). Information on the interior, such as lithosphere–mantle–core configurations, can be obtained from instruments, such as seismometers, or from geophysical models based on knowledge of the planetary density, size, moment of inertia and shape. Information on the exterior environment includes the temperature range, mechanical and thermal properties of the surface materials, the influence of various external processes (such as impact cratering), solar wind flux and the presence or absence of an atmosphere and, if present, its composition and density. Finally, knowledge is required of the various active processes which may be operating on the planetary object, such as volcanism.

1.2.2 Past state of the planets

This objective is to trace the processes, events and characteristics of the planets from their origin to the present; in other words, to determine their geologic histories. The approach for meeting this goal is met primarily through geologic mapping, in which surface materials (rock units) are identified and placed in a time sequence. In planetary geology, mapping is accomplished primarily by photogeologic techniques; however, as we shall see later, this method is not without some problems. Despite the difficulties, the level of mapping which is attainable at least provides a broad framework for the derivation of geologic histories.

1.2.3 Comparative planetology

Once knowledge is gained—even if incomplete—of the present and past states of the planets, it is then possible to begin to compare the planets to see how they are similar to and different from each other. Such comparisons help to meet the objectives of Solar System exploration by enabling a better understanding of the evolution of all

Figure 1.3 Apollo 11 astronaut on Moon, showing various experiments deployed around the lander. Samples returned from the Moon by the Apollo 11 crew enabled for the first time detailed compositional determination and dating of the ages of emplacement of mare lavas (Apollo 11 69-HC-898).

the objects in the Solar System and of the processes involved in their formation and evolution.

1.3 Relevance of geomorphology

Ideally, the various geologic objectives of Solar System exploration would be met by carrying out the entire sequence of missions and measurements, as indicated in Table 1.1. Limited resources prevent the completion of this strategy for Solar System exploration, at least in the foreseeable future. Meanwhile, nearly all previous and near-term anticipated missions have been equipped with **imaging systems** to acquire pictures of planetary surfaces. Aside from the obvious appeal of pictures, it is generally recognized that imaging science can yield answers to a broader range of questions than any other single instrument which might be carried on board spacecraft.

Planetary pictures also play a key role from the standpoint of engineering. During mission operations, flight engineers use pictures taken by the spacecraft for celestial navigation in guiding the probe in its journey through space. For missions involving landers, pictures of potential landing sites play a key role in the selection of safe sites.

Figure 1.5 Apollo 15 photograph of the Service Module taken from the Landing Module, showing the Scientific Instrument Module (SIM) containing various remote-sensing experiments, including cameras, alpha, gamma-ray and X-ray spectrometers (Apollo 15 AS 15-88-11972).

Figure 1.4 View of the martian surface from Viking Lander 1, showing meteorology boom (left side, extending out of view toward top) used to measure wind speed and direction, temperature and atmospheric pressure, and sample arm (lower middle) used to obtain samples for analysis. Arm could dig into surface (trench for sample acquisition is marked with arrow) to retrieve the sample and dump it into a hopper on the spacecraft for analysis.

The first phase of most planetary geologic studies involves the classification of various landforms observed through images and the production of **physiographic maps** (Fig. 1.6). The next step is to derive geologic maps using various photogeologic techniques. Color, **albedo** (reflective properties of the surface), texture and other remotely sensed characteristics are used to distinguish possible rock formations. Then, using various geometric relationships, such as superposition, embayment and cross-cutting relations, the identified formations are placed in a relative sequence. These techniques have been long established in photogeologic studies of the Earth. An additional technique useful in planetary geology for relative dating of some surfaces is to establish the size–frequency distribution of impact craters. The idea is that surfaces act as impact "counters": the older the surface, the higher the frequency of superposed craters (Fig. 1.7). This concept will be discussed in more detail in Chapter 3.

Figure 1.6 Diagrams showing types of maps commonly used to portray planetary surfaces (from Spudis & Greeley 1976). (a) *Geologic maps* show three-dimensional rock units in space and time, plus structural features such as folds and faults. They portray the surface of the planet as though all vegetation, soil and other surficial materials were stripped away; sequences of geologic events can be derived from this type of map. (b) *Physiographic maps* show landforms such as hills, valleys and plains. (c) *Surficial (soil) maps* show the distribution only of surficial covers or the lack thereof and do not show topography or terrain types. (d) *Surficial geology maps* which characterize the local geologic and surficial geology of a given area. Note that this map presents all the data of the physiographic (terrain) and surficial (soil) maps, but with a substantial reduction in map complexity.

Figure 1.7(a) Sequence of photographs from laboratory simulations showing evolution of a cratered surface, beginning in upper left with smooth, uncratered surface. As time progressed, the total numbers of craters increased and statistically there was a greater chance for large craters to form. Thus, by comparing the size-frequency distribution of craters among different surfaces, it is possible to obtain relative dates for those surfaces. For example, frame 3 is sparsely cratered in comparison to frame 10, and is therefore younger. Note, however, that the last six frames all appear to be about the same; this surface is said to be *in equilibrium*, in which craters of a given size are being destroyed at the same rate of formation (NASA-Ames photograph AAA-312-23, from Gault 1970).

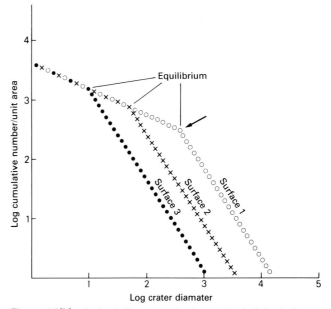

Figure 1.7(b) Stylized diagram indicating method of displaying crater statistics in which cumulative numbers of craters in given size ranges are plotted per unit surface area. "Break" in slope (arrow) marks upper crater size that has reached "equilibrium" (see Gault 1970). As a surface "ages", the total number of craters shifts toward larger sizes and greater numbers; surface 1 (oldest), surface 3 (youngest), assuming all craters are of impact origin.

A true geologic map should show the distribution of *material* units, that is, three-dimensional rock units, or *formations*. The test of a geologic map is to draw a cross-section through the map; the units should appear to have a finite thickness (Fig. 1.6). In mapping some planetary surfaces, often it is not possible to construct true geologic maps owing to lack of adequate data; rather, the maps show *surficial geology* or are some combination of geologic and terrain maps as shown in Figure 1.6.

Once features and units are classified and tentatively identified, interpretations are attempted to determine the processes involved in their formation. This is often the most difficult and controversial part of planetary geology. Typically, a three-fold approach is used—results from spacecraft data analysis, modeling and studies of terres-

9

trial analogs are combined to derive the best possible answers for the problem at hand. Figure 1.8 illustrates how this approach has been used in planetology to understand the process of impact cratering.

First, the problem is defined from the study of spacecraft images; in the case illustrated here (Fig. 1.8(a)), the objective is to understand impact cratering as a geologic process. Next, impact cratering is physically modeled in the laboratory where the various parameters, such as the velocity of the impacting object, can be controlled, isolated and studied to determine their effects in the process (Figs. 1.8(b), (c) and (d)). Numerical modeling is also carried out, based on theory, to fill in detail and to gain insight into parts of the problem that are difficult to simulate in the laboratory. In general, physical and numerical modeling are first carried out for environments appropriate for conditions on Earth. Then, natural impact craters on Earth (Fig. 1.8(e)) are studied geologically as terrestrial analogs to the features observed on planetary surfaces. These field studies serve as important checks on the modeling results and provide direct information on the full-scale, complex process. Results from the field studies then can be used to modify, correct or adjust the modeling techniques. Once confidence is established in the ability to model the terrestrial case, the simulations can be carried out for the planetary case under study, using values for parameters, such as gravity, that are appropriate for the planet involved. The final results can then be applied to the interpretation of the planetary problem first defined, thus completing the study cycle.

The approach outlined above requires a multidisciplinary team effort, because any one individual seldom has the combined talents and background of geology, chemistry, physics and other fields to handle all aspects of the problem. The ultimate result of the approach is not only the solution of planetary problems but also an understanding of various processes from a "universal" perspective, regardless of planetary object.

Planetologists may not think of this approach in geomorphic terms, but the underlying theme of much of the field is, in fact, geomorphology—be it in terrain mapping or gaining an understanding of surface-forming and surface-modifying processes. Sharp (1980) and Baker (1984) provide summaries of the role of geomorphology in the study of planetary surfaces.

1.4 Sources of data

The results from Solar System exploration are presented in nearly all scientific journals and popularized serials. However, geologic results are more commonly found in a relatively restricted set of publications. Preliminary

Figure 1.8(a) Impact craters occur on nearly all solid-surface objects that have been observed in the Solar System. Understanding the mechanics of impact cratering and the geomorphic characteristics in terms of geologic processes are of key importance in planetary science. This view, obtained by the Apollo 15 astronauts from orbit, shows the craters Autolycus (nearest crater) and Aristillus, a 50 km diameter crater having a central peak; both craters show enlargement by slumping of the walls (Apollo 15 metric frame AS 15-1539).

Figure 1.8(b) View of the Vertical Ballistic Gun (VBG) facility at NASA-Ames Research Center used for impact cratering experiments. Projectiles can be fired down the gun tube at velocities up to 7.5 km s^{-1} into a vacuum tank to impact targets. Gun tube can be elevated in 15° increments to enable crater experiment to be conducted for a range of incidence angles. High-speed motion pictures (exceeding 10^6 frames per second) obtained during the cratering event enables analysis of cratering sequence (NASA-Ames photograph A33996, courtesy of D. E. Gault).

Figure 1.8(c) Photograph inside chamber of VBG showing typical experiment; crater 40 cm in diameter was formed by a 6 mm glass projectile launched at 6.35 km s^{-1} at an incidence angle of 45° (entering from the right); note the circular crater form and distribution of ejecta (throw-out from crater) despite the oblique impact angle.

Figure 1.8(d) Cross section of impact crater in VBG formed in non-cohesive target of quartz sand with colored layers, revealing overturned strata in the rims and compressed strata beneath the crater (from Gault 1974).

Figure 1.8(e) Oblique aerial photograph of Meteor Crater in northern Arizona, showing eroded ejecta deposits (mottled patterns around crater and blocks on the rim and flanks) and polygonal outline which resulted from pre-crater joints in the target. Field studies of the well preserved 1.2 km diameter crater provide important "checks" on results from cratering simulations and contribute to the overall knowledge of the cratering process (photograph by Malin 1976).

results from US missions are commonly published in *Science*. Subsequent results typically are published later in journals such as *Icarus, Nature*, the *Proceedings of the Lunar and Planetary Science Conference, Earth, Moon and Planets* (formerly *The Moon*), *Earth and Planetary Science Letters* and the *Journal of Geophysical Research*. Table 1.2 lists some of the more important issues devoted to results of planetary exploration. Books introducing planetary geology include King (1976), Guest *et al.* (1979), Murray *et al.* (1981), Beatty and Chaikin (1990), Glass (1982), Taylor (1982), Hartmann (1983), and Hamblin and Christiansen (1990).

Although many scientific meetings are held each year for presentations and discussions of planetary geology research, the key international meeting is the Lunar and Planetary Science Conference held each spring at the NASA-Johnson Space Center, Houston. An abstract volume (available from the Lunar and Planetary Institute; Table 1.4) is published each year for the conference. Typically exceeding 1000 pages in length, the volume gives summaries of results and works in progress. In addition, an issue of *Geotimes* (late spring or summer) is devoted to results from the conference and provides a quick means for following progress in planetary geology.

NASA publishes planetary results through several series and often produces collections of photographs from various missions (Table 1.3). Lunar and planetary maps and charts have been produced by both the US Department of Defense and the US Geological Survey. The Planetary Data Facility, US Geological Survey, Flagstaff, Arizona, provides up-to-date listings of available maps and charts for planets and satellites.

Images from United States' lunar and planetary missions are available from the National Space Sciences Data Center (requests from within the United States) and the World Data Center (requests from outside the United States), at Goddard Spaceflight Center, NASA, Greenbelt, Maryland, USA; both will supply catalogs of images upon request. In addition, various centers (Table 1.4) have been established to provide access to planetary images by serious investigators.

In the chapters that follow, we will first consider the Solar System in the geologic context and review the processes involved in the formation and evolution of the planets. The physiography of each planet and satellite of geologic importance is then presented in separate chapters.

Table 1.2 Special journal issues containing planetary geology references.

Planet	Mission	Journal
Mercury	Mariner 10	*Science,* 1974, **185**, no. 4146
Mercury	Mariner 10	*J. Geophys. Res.,* 1975, **80**, no. 17
Mercury	Mariner 10	*Phys. Earth Planet. Int.,* 1977, **15**, nos. 2 and 3
Mercury	Mariner 10	*Icarus,* 1976, **28**, no. 4
Venus	Mariner 10	*Science,* 1974, **183**, no. 4131
Venus	Pioneer	*Science,* 1979, **203**, no. 4382
Venus	Pioneer	*Science,* 1979, **205**, no. 4401
Venus	Pioneer	*J. Geophys. Res.,* 1980, **85**, no. A13
Venus	general	*Icarus,* 1982, **51**, no. 2
Venus	general	*Icarus,* 1982, **52**, no. 2
Venus	Vega	*Science,* 1986, **231**, no. 4744
Venus	Magellan	*Science,* 1991, **252**, no. 5003
Venus	Magellan	*J. Geophys. Res.,* 1992, **97**, nos. E8 and E9
Moon	Apollo II	*Science,* 1970, **167**, no. 3918
Moon	Apollo	*The Moon,* 1974, **9**
Moon	general	*Rev. Geophys. Space Phys.,* 1974, **12**, no. 1
Moon	general	*The Moon,* 1975, **13**, nos. 1, 2 and 3
Moon	Galileo	*Science,* 1992, **255**, no. 5044
Mars	Mariner 6 and 7	*J. Geophys. Res.,* 1971, **76**, no. 2
Mars	Mariner 9	*Icarus,* 1972, **17**, no. 2
Mars	Mariner 9	*Icarus,* 1973, **18**, no. 1
Mars	Mariner 9	*J. Geophys. Res.,* 1973, **78**, no. 20
Mars	Mariner 9	*Icarus,* 1974, **22**, no. 3
Mars	Viking 1	*Science,* 1976, **193**, no. 4255
Mars	Viking 1 and 2	*Science,* 1976, **194**, no. 4260
Mars	Viking	*J. Geophys. Res.,* 1977, **82**, no. 28
Mars	Viking	*Icarus,* 1978, **34**, no. 3
Mars	general	*J. Geophys. Res.,* 1979, **84**, no. B14
Mars	general	*Icarus, 1981,* **45**, nos. 1 and 2
Mars	general	*Icarus,* 1982, **50**, nos. 2 and 3
Mars	general	*J. Geophys. Res.,* 1982, **87**, no. B12
Mars	general	*J. Geophys. Res.,* 1990, **95**, no. B9
Jupiter	Pioneer 11	*Science,* 1975, **188**, no. 4187
Jupiter	Voyager 1	*Nature,* 1979, **280**, no. 5725
Jupiter	Voyager 1	*Science,* 1979, **204**, no. 4396
Jupiter	Voyager 2	*Science,* 1979, **206**, no. 4421
Jupiter	Voyager	*J. Geophys. Res.,* 1981, **86**, no. A10
Saturn	Pioneer 11	*Science,* 1980, **207**, no. 4429
Saturn	Voyager 1	*Nature,* 1981, **292**, no. 5825
Saturn	Voyager 1	*Science,* 1981, **212**, no. 4491
Saturn	Voyager 2	*Science,* 1982, **215**, no. 4532
Saturn	Voyager	*Icarus,* 1983, **53**, no. 2
Uranus	Voyager 2	*Science,* 1986, **233**, no. 4739
Neptune	Voyager 2	*Science,* 1989, **246**, no. 4936
Neptune	Voyager 2	*J. Geophys. Res.,* 1991, **96**, supplement
Neptune-Triton	Voyager 2	*Science,* 1990, **250**, no. 4979
Comet Halley	5 missions	*Nature,* 1986

Table 1.3 NASA publications of relevance to planetary geology.

Object	Year	Serial	Title	Notes
Moon	1964	SP-61	*Ranger VII photographs of the Moon,* Part I, Camera A series	
Moon	1965	SP-62	*Ranger VII photographs of the Moon,* Camera B series	Photographic collection
Moon	1965	SP-63	*Ranger VII photographs of the Moon,* Camera P series	Photographic collection

Table 1.3 *Continued*

Object	Year	Serial	Title	Notes
Moon	1966	SP-111	*Ranger VIII photographs of the Moon,* Cameras A,B,P	Photographic collection
Moon	1966	SP-112	*Ranger IX photographs of the Moon*	Photographic collection
Moon	1966	SP-126	*Surveyor I: a preliminary report*	
Moon	1966	JPL-TR 32-800	*Ranger VIII and IX experimenters' analysis and interpretations*	Mission description and science results
Moon	1969	SP-184	*Surveyor: program results*	
Moon	1969	SP-214	*Apollo 11 preliminary science report*	Mission description with photographs;
Moon	1969	SP-201	*Apollo 8, photography and visual observations*	includes descriptions and results from various experiments on board the command module and/or lander spacecraft
Moon	1970	SP-242	*Guide to Lunar orbiter photographs*	Explains camera system and gives footprints
Moon	1970	SP-200	*The Moon as viewed by Lunar orbiter*	Mission description and photographs
Moon	1970	SP-235	*Apollo 12 preliminary science report*	Mission description with photographs; includes descriptions and results from various experiments on board the command module and/or lander spacecraft
Moon	1971	SP-241	*Atlas and gazetteer of the near side of the Moon*	Lunar orbiter photographs with place names
Moon	1971	SP-206	*Lunar orbiter photographic atlas of the Moon*	Photographic collection
Moon	1971	SP-232	*Apollo 10, photography and visual observations*	Mission description with photographs;
Moon	1971	SP-238	*Apollo 11 mission report*	includes descriptions and results
Moon	1971	SP-272	*Apollo 14 preliminary science report*	from various experiments on board the
Moon	1971	SP-246	*Lunar photographs from Apollos 8, 10, 11*	command module and/or lander spacecraft
Moon	1972	SP-315	*Apollo 16 preliminary science report*	
Moon	1972	SP-289	*Apollo 15 preliminary science report*	
Moon	1972	SP-284	*Analysis of Surveyor 3 material and photographs returned by Apollo 12*	
Moon	1972	SP-306	*Compositions of major and minor minerals in five Apollo 12 crystalline rocks*	Description of samples
Moon	1973	SP-330	*Apollo 17 preliminary science report*	Mission description with photographs; includes descriptions and results from various experiments on board the command module and/or lander spacecraft

Table 1.3 *Continued*

Object	Year	Serial	Title	Notes
Moon	1974	EP-100	*Apollo*	Public information booklet
Moon	1973	SP-341	*Atlas of Surveyor 5 television data*	Photographs with short captions
Moon	1975	SP-350	*Apollo expedition to the Moon*	Mission description with photographs, general public
Moon	1977	SP-418	*Lunar sample studies*	
Moon	1978	SP-362	*Apollo over the Moon*	Summary of Apollo missions to the Moon; color photographs, science discussions
Mars	1968	SP-179	*The book of Mars*	
Mars	1971	SP-263	*The Mariner 6 and 7 mission to Mars*	Mission description and photographic collection
Mars	1974	SP-334	*The Viking mission to Mars*	
Mars	1974	SP-329	*Mars as viewed by Mariner 9*	Photographs with science
Mars	1974	SP-337	*The new Mars, the discovery of Mariner 9*	Photographs with science
Mars	1978	SP-425	*The martian landscape*	Mission description and photographs for Viking landers
Mars	1979	SP-438	*Atlas of Mars, the 1:5000000 map series*	Map and photomosaic collection
Mars	1980	SP-444	*Images of Mars—the Viking extended mission*	Photograph collection
Mars	1980	SP-441	*Viking orbiter views of Mars*	Photo collection with science
Mars	1980	CR-3326	*The mosaics of Mars as seen by the Viking lander cameras*	Lander photomosaics with discussion of assembly
Mars	1981	SP-429	*Viking site selection and certification*	
Mars	1982	CR-3568	*Viking lander atlas of Mars*	Lander photographs and maps
Mars	1983	RP-1093	*A catalog of selected Viking orbiter images*	Photomosaics based on Mars charts; gives frame locations
Mars	1984	SP-4212	*On Mars, exploration of the red planet 1958-1978*	History of missions and exploration
Mercury	1978	SP-423	*Atlas of Mercury*	Synopsis of Mariner 10 results, collection of photographs and USGS charts
Mercury	1978	SP-424	*The voyage of Mariner 10*	Mission description
Venus	1975	SP-382	*The atmosphere of Venus*	
Venus	1983	SP-461	*Pioneer Venus*	Popularized account of mission with discussion of science results
Jupiter	1971	SP-268	*The Pioneer mission to Jupiter*	
Jupiter	1974	SP-349	*Pioneer Odyssey— encounter with a giant*	Mission description and photographic collection
Jupiter	1980	SP-439	*Voyage to Jupiter*	Popularized account of mission with discussion of science results
Jupiter	1989	SP-494	*Time-variable phenomena in the jovian system*	

Table 1.3 *Continued*

Object	Year	Serial	Title	Notes
Jupiter/ Saturn	1977	SP-420	*Voyager to Jupiter and Saturn*	
Jupiter/ Saturn	1980	SP-446	*Pioneer: first to Jupiter, Saturn, and beyond*	
Saturn	1974	SP-340	*The atmosphere of Titan*	
Saturn	1974	SP-343	*The rings of Saturn*	
Saturn	1980	JPL-400-100	*Voyager 1 encounters Saturn*	Public information booklet
Saturn	1982	SP-451	*Voyages to Saturn*	Popularized account of mission with discussion of science results
Saturn	1984	SP-474	*Voyager 1 and 2 atlas of six saturnian satellites*	
general	1971	SP-267	*Physical studies of the minor planets*	
general	1976	SP-345	*Evolution of the Solar System*	
general	1981	EP-177	*A meeting with the universe*	Popularized account of Solar System exploration and other space science activities
general	1984	SP-469	*The geology of the terrestrial planets*	An introduction to Mercury, Venus, Earth, Moon and Mars, and a chapter on asteroids, comets and planetary formation

Table 1.4 Regional and branch space image facilities.

Brown Regional Planetary Data Center, Brown University, Providence, RI 02912

Space Imagery Center, Lunar and Planetary Lab, University of Arizona, Tucson, AZ 85721

Spacecraft Planetary Imaging Facility, Cornell University, 317 Space Sciences Bldg, Ithaca, NY 14853

Planetary Image Facility, Jet Propulsion Laboratory, Bldg 264, Rm 115, 4800 Oak Grove Drive, Pasadena, CA 91103

Planetary Image Facility, Lunar and Planetary Institute, 3600 Bay Area Blvd, Houston, TX 77058

Flagstaff Planetary Data Facility, USGS, 2255 N Gemini Drive, Flagstaff, AZ 86001

Regional Planetary Image Facility, Department of Earth and Planetary Science, Washington University, St Louis, MO 63130

Regional Planetary Image Facility, Smithsonian Institute, Washington, DC 20560

Space Photography Laboratory,[*] Department of Geology, Arizona State University, Tempe, AZ 85287

Regional Planetary Image Facility, University of Hawaii, Hawaiian Institute of Geophysics, Honolulu, HI 96822

Regional Space Image Library, University of London Observatory, Mill Hill Park, 2QS SW7, UK

Fotote La-archivio Immagini E Dati Planeteri, Reparto Planetogia, Viale Universite 11-00185, Roma, Italy

Space Image Library, Laboratoire de Géologie dynamique interne (Bat 509), University of Paris Sud, 91405 Orsay, France

Planetary Image Facility, ISAS, Division of Planetary Science, Kanagawa 229, Japan

Space Image Library, Deutsche Forschungsanstalt für Luft- und Raumfahrt e.V., Institute for Planetary Exploration, Reginoal Planetary Image Facility, Rudower Chaussee 5, O-1199 Berlin, Germany

[*]Designates branch facilities not generally open to the public.

14

2 Geologic exploration of the Solar System

Earth-based telescopic observations of the planets and satellites began in the early 17th century. Through the years these observations have enabled some of the general properties of the planets, such as sizes, densities and orbital characteristics, to be determined. Space probes have added substantially to this knowledge, particularly in regard to the outer planet satellites—some of which were unknown prior to their discovery on spacecraft images. In this chapter we will examine the general properties of the planets and satellites important for geologic analyses. We will also consider the types of space probes involved in Solar System exploration and discuss the means by which geologic data—primarily images—are acquired.

Spacecraft images are extremely important to planetary geology because they are the primary source for understanding the geomorphology of the planets and satellites. Unless the researcher is familiar with how the images are acquired and processed, spacecraft pictures can easily be misinterpreted. Thus, in the last section of this chapter, we will discuss camera systems that have been utilized on various missions along with the techniques that are employed for the analysis of spacecraft images.

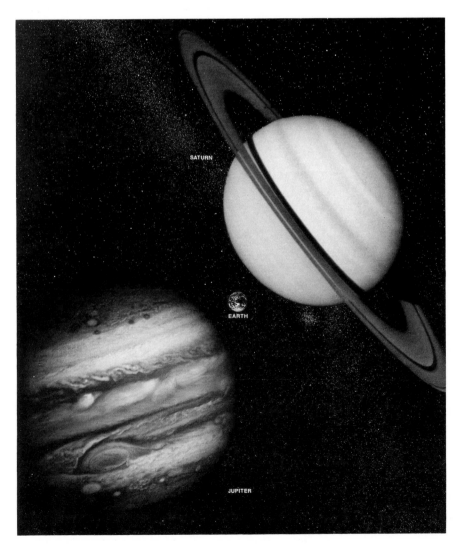

Figure 2.1 Comparison of terrestrial planet (Earth) with two jovian planets (Saturn and Jupiter) all shown to same scale (NASA photograph 83 H 202).

2.1 General planetary characteristics

Astronomers have long recognized that the planets can be classified into two groups, the **terrestrial planets** and the **jovian planets**. As the names imply, terrestrial planets are "Earthlike"; these are relatively small, dense objects, composed mainly of silicate materials that have solid surfaces (Fig. 2.1). The terrestrial planets, also called the inner planets from their position in the Solar System, are Mercury, Venus, Earth and Mars. Most investigators consider Earth's Moon to be a terrestrial planet because of its large size and high density (Table 2.1).

Table 2.1(a) Basic data for planets.

Name	Orbit Semi-Major Axis (10^6 km)	(AU)	Revolution Period	Diameter (km)	Rotation[*] (days)	Mass (10^{20} kg)	Density (g/cm^3)	Escape Velocity (km/s)	Surface	Atmosphere
Mercury	57.9	0.39	0.24 yr	4878	58.65	3.3	5.4	4	Silicates	Trace Na
Venus	108	0.72	0.62 yr	12,102	243.0 R	48.7	5.3	10	Basalt, granite?	90 bar: 97% CO_2
Earth	150	1.00	1.00 yr	12,756	1.00	59.8	5.5	11	Basalt, granite, water	1 bar: 78% N_2, 21% O_2
Mars	228	1.52	1.88 yr	6787	1.03	6.4	3.9	5	Basalt, clays, ice	0.07 bar: 95% CO_2
Jupiter	778	5.20	11.86 yr	142,984	0.41	18,991	1.3	60	None	H_2, He, CH_4, NH_3, etc.
Saturn	1426	9.54	29.46 yr	120,536	0.44	5,686	0.7	36	None	H_2, He, CH_4, NH_3, etc.
Uranus	2868	19.18	84.07 yr	51,118	0.72 R	866	1.2	21	None (?)	H_2, He, CH_4, NH_3, etc.
Neptune	4494	30.06	164.82 yr	49,244	0.7	1,030	1.6	23	None (?)	H_2, He, CH_4, NH_3, etc.
Pluto	5900	39.44	247.7 yr	2400	6.39 R	0.01	2.0	1	CH_4 ice	Trace CH_4

[*]R = retrograde

Table 2.1(b) Basic data for satellites.

Planet	Satellite Name	Discovery	Semi-major Axis (km × 1000)	Period (days)	Diameter (km)	Mass (10^{20} kg)	Density (g/cm^3)	Surface Material
Earth	Moon	—	384	27.32	3,476	735	3.3	silicates
Mars	Phobos	Hall (1877)	9.4	0.32	27	1×10^{-4}	2.2	carbonaceous
	Deimos	Hall (1877)	23.5	1.26	13	2×10^{-5}	1.7	carbonaceous
Jupiter	Metis	Voyager (1979)	128	0.29	40	—	—	rock?
	Adrastea	Voyager (1979)	129	0.30	25	—	—	rock?
	Amalthea	Barnard (1892)	181	0.50	270	—	—	rock, sulfur
	Thebe	Voyager (1979)	222	0.67	110	—	—	rock?
	Io	Galileo (1610)	422	1.77	3,630	889	3.6	sulfur, SO_2
	Europa	Galileo (1610)	671	3.55	3,138	479	3.0	ice
	Ganymede	Galileo (1610)	1,070	7.15	5,262	1,490	1.9	dirty ice
	Callisto	Galileo (1610)	1,883	16.69	4,800	1,064	1.8	dirty ice
	Leda	Kowal (1974)	11,094	239	16	—	—	?
	Himalia	Perrine (1904)	11,480	251	186	—	—	carbonaceous
	Lysithea	Nicholson (1938)	11,720	259	36	—	—	carbonaceous
	Elara	Perrine (1905)	11,737	260	76	—	—	carbonaceous
	Ananke	Nicholson (1951)	21,200	631 (R)[a]	30	—	—	?
	Carme	Nicholson (1938)	22,600	692 (R)	40	—	—	?
	Pasiphae	Melotte (1908)	23,500	735 (R)	50	—	—	?
	Sinope	Nicholson (1914)	23,700	758 (R)	36	—	—	?
Saturn	Pan	Voyager (1985)	133.6	0.58	20?	3×10^{-5}	—	?
	Atlas	Voyager (1980)	137.7	0.60	40	—	—	ice?
	Prometheus	Voyager (1980)	139.4	0.61	140	—	—	ice?
	Pandora	Voyager (1980)	141.7	0.63	110	—	—	ice?
	Janus	Dollfus (1966)	151.4	0.69	220	—	—	ice?
	Epimetheus	Fountain, Larson (1980)	151.4	0.69	140	—	—	ice?
	Mimas	Herschel (1789)	186	0.94	392	0.376	1.2	ice
	Enceladus	Herschel (1789)	238	1.37	500	0.740	1.1	pure ice
	Tethys	Cassini (1684)	295	1.89	1,060	6.26	1.0	ice
	Telesto	Reitsema et al. (1980)	295	1.89	34	—	—	ice?
	Calypso	Pascu et al. (1980)	295	1.89	34	—	—	ice?
	Dione	Cassini (1684)	377	2.74	1,120	11	1.4	ice
	Helen	Lecacheux, Laques (1980)	377.4	2.74	36	—	—	ice?
	Rhea	Cassini (1672)	527	4.52	1,530	22	1.3	ice

Table 2.1(b) *Continued*

Planet	Satellite Name	Discovery	Semi-major Axis (km × 1000)	Period (days)	Diameter (km)	Mass (10^{20}kg)	Density (g/cm^3)	Surface Material
Saturn	Titan	Huygens (1655)	1,222	15.95	5,150	1,346	1.9	cloudy atmosphere
(cont'd)	Hyperion	Bond, Lassell (1848)	1,481	21.3	410	—	—	dirty ice
	Iapetus	Cassini (1671)	3,561	79.3	1,460	19	1.1	ice/carbonaceous
	Phoebe	Pickering (1898)	12,952	550 (R)	220	—	—	carbonaceous?
Uranus	Cordelia	Voyager (1986)	49.7	0.34	26	—	—	?
	Ophelia	Voyager (1986)	53.8	0.38	30	—	—	?
	Bianca	Voyager (1986)	59.2	0.43	42	—	—	?
	Cressida	Voyager (1986)	61.8	0.46	62	—	—	?
	Desdemona	Voyager (1986)	62.7	0.47	54	—	—	?
	Juliet	Voyager (1986)	64.4	0.50	84	—	—	?
	Portia	Voyager (1986)	66.1	0.51	108	—	—	?
	Rosalind	Voyager (1986)	69.9	0.56	54	—	—	?
	Belinda	Voyager (1986)	75.3	0.62	66	—	—	?
	Puck	Voyager (1985)	86.0	0.76	154	—	—	carbonaceous?
	Miranda	Kuiper (1948)	129.3	1.41	480	0.2	1.3	dirty ice
	Ariel	Lassell (1851)	191	2.52	1,158	16	1.6	dirty ice
	Umbriel	Lassell (1851)	266.3	4.14	1,172	9.3	1.4	dirty ice
	Titania	Herschel (1787)	436	8.71	1,580	28	1.6	dirty ice
	Oberon	Herschel (1787)	583	13.5	1,524	29	1.5	dirty ice
Neptune	Naiad	Voyager (1989)	48	0.30	54	—	—	—
	Thalassa	Voyager (1989)	50	0.31	80	—	—	—
	Despina	Voyager (1989)	53	0.33	150	—	—	—
	Galatea	Voyager (1989)	62	0.43	160	—	—	—
	Larissa	Voyager (1989)	74	0.55	208	—	—	—
	Proteus	Voyager (1989)	118	1.12	436	—	—	—
	Triton	Lassell (1846)	335	5.88(R)	2,700	—	2.0	methane ice
	Nereid	Kuiper (1949)	5,513	360	340	—	—	dirty ice
Pluto	Charon	Christy (1978)	19.6	6.39	1,186	—	—	ice

[a]R = retrograde.

Jove is another name for the Roman god Jupiter; thus, jovian planets share the characteristics of the planet Jupiter (Fig. 2.1). They are large, low-density objects composed mostly of hydrogen and helium. They all apparently lack solid surfaces and, thus, are not amenable to geologic study. The jovian planets are Jupiter, Saturn, Uranus and Neptune. These are also known as the **outer planets** by their position in the Solar System. Pluto is also an outer planet, but its small size precludes its inclusion with the jovian planets. Although the outer planets are not of direct geologic interest, their satellites are solid-surface objects and thus can be studied in the geologic context.

Observations of the surfaces of the Martian moons Phobos (~22 km in diameter) and Deimos (~14 km in diameter) and the asteroid, Gaspra, introduced a new class of object to planetary geomorphology. Although not defined by a specific size, this class is generally referred to as **small bodies** and includes comets and the smaller satellites of the outer planets.

The satellites of Jupiter, Saturn, Uranus and Neptune display a wide range of sizes and characteristics. The Galilean satellites of Jupiter (so named after their discoverer, Galileo Galilei, in the early 1600s) are respectable-size objects in their own right. The two innermost moons,

Io and Europa, are comparable in size and density to Earth's Moon and might also be considered as terrestrial planets. The outer two Galilean satellites are Ganymede and Callisto. They are roughly the size of Mercury, but their low density suggests that they are composed of a mixture of water and silicates.

2.2 Pre-space-age planetology studies

Although the first telescopes used to view the Moon and planets were rather crude, technology advanced rapidly and by the mid-to-late 1600s maps of the Moon based on telescopic observations were remarkably detailed. In 1651, Giovanni Riccioli, a Jesuit priest, made a comprehensive map of the nearside of the Moon—the side that always faces Earth. The smooth, flat, dark areas were interpreted to be bodies of water and were named accordingly as **maria** (Fig. 1.2), the Latin word for seas; the large craters are named after people, such as Copernicus and Kepler. Hevelius, a German astronomer, also mapped the Moon and named lunar mountain chains after moun-

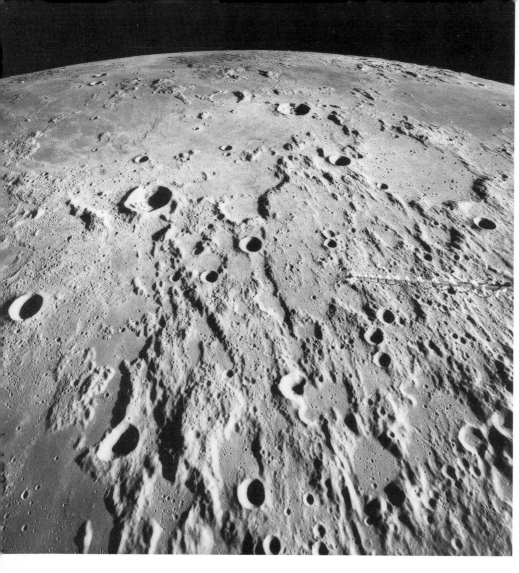

Figure 2.2 View northward obtained by the Apollo 16 astronauts showing Imbrium sculpture northwest of the crater Ptolemaeus. Sculpture resulted from gouging of ejecta from the Imbrium basin, visible at top of photograph (Apollo 16 AS 16-1412).

tain ranges on the Earth, so that today we find the lunar Alps and Apennines on modern maps of the Moon.

With the development of photography, the lunar nearside was extensively documented at resolutions as good as a few kilometers. The variety of lunar surface features that were discovered through observations and photographs sparked considerable controversy as to their origin and much of this controversy was not laid to rest until well into the space age.

Craters are the most prominent surface features on the Moon and have long generated curiosity as to their origin. Green (1965) provides an interesting historical account of the controversy surrounding lunar craters. He notes that the American geologist G. K. Gilbert was one of the first investigators to consider lunar craters in the geologic context and supported earlier speculations that the craters were of impact origin. In the late 1800s, he recognized surface textures and radial patterns around Mare Imbrium (Fig. 2.2) and interpreted the features to be the result of gouging of the surface by material ejected from a huge impact event (Gilbert 1893). Gilbert also carried out ex-

periments in attempts to simulate the impact cratering process (Fig. 2.3).

The 1940s to early 1960s saw the development of ideas to explain not only lunar craters but also other aspects of the general geology of the Moon. Ralph Baldwin's books (1949 and 1963) and the works of Harold Urey (1952), Gerard Kuiper and Robert Dietz described a wide variety of surface features; Spurr published a series of books from 1944 to 1949 (Spurr 1944, 1945, 1948 and 1949) detailing various aspects of lunar geology. His work was followed by that of Gilbert Fielder, an English astronomer who attributed most lunar features to volcanism. Although Fielder's (1961, 1965) ideas on volcanic origins for lunar craters are not now accepted, many of his hypotheses regarding lunar lava flows and possible volcanic domes appear to be correct.

With the development of the United States' space program in the early 1960s came a commitment to carry out scientific studies of the Moon and planets. Primarily in support of the Apollo program, the Branch of Astrogeology was created as part of the US Geological Survey.

Figure 2.3 Photograph by G. K. Gilbert taken *c.* 1891 showing his impact cratering experiments in which balls of clay impacted a slab of clay; he noted that the crater morphology was dependent upon the impact velocity (US Geological Survey photograph, G. K. Gilbert No. 842).

Among other responsibilities, this organization was charged with the task of mapping the geology of the Moon from Earth-based observations. Eugene Shoemaker and R. J. Hackman (1962) demonstrated the feasibility of making geologic maps of the Moon. It was recognized that such maps would play a key role in selecting future landing sites and for plotting the scientific results of lunar exploration within a logical framework. This work was considered so important that an observatory was constructed near Flagstaff, Arizona—home of the Branch of Astrogeology—to provide unhindered observation time for lunar geologic mapping.

Aside from the Moon, Mars is the only other planetary object which can be observed from Earth with sufficient resolution to study surface features. Descriptions of canals on Mars have appeared in writings for more than 100 years, but the canals failed to appear on high-resolution pictures of Mars obtained by spacecraft and are now considered to have been figments of the observers' imaginations. However, Mars displays various surface markings and atmospheric features which appear regularly in concert with the Martian seasons. In the mid-1950s Dean McLaughlin speculated that these features, long observed telescopically, might be the result of dust storms. He wrote a series of papers (1954) documenting these and other phenomena on Mars and even published a map showing wind circulation patterns based on tracking of certain albedo patterns now known to be clouds of dust. These and similar features have been seen on hundreds of spacecraft images and many of the ideas proposed by McLaughlin regarding wind processes appear to be valid.

2.3 Lunar and planetary missions

Table 2.2 lists the important "firsts" in spacecraft missions of geologic relevance and includes the first lunar flyby of the Soviets in 1959. This was the first of several series of unmanned missions designed to obtain critical engineering and scientific data.

Ranger spacecraft, first of the unmanned US lunar missions to return geologic data, were launched on trajectories that carried them to direct impact on the Moon. After a series of disappointing failures, Ranger 7 successfully obtained the first close-up pictures of the Moon in the summer of 1964; Rangers 8 and 9 followed a short time later and again returned high-resolution images. In flight these spacecraft took a continuous stream of television pictures of progressively higher resolution as the probes approached the surface, right to the moment of impact. Having a spatial resolution of about 1 m, these last pictures showed that there was a continuum of crater sizes down to the limit of resolution (Fig. 2.4). Analysis of these images showed that the number of progressively smaller craters increases exponentially.

The year 1966 marked the beginning of two important series of US unmanned spacecraft, the **Surveyor** series and the **Lunar orbiter** series. As the name implies, Lunar orbiters were placed in orbit around the Moon and had the principal objective of obtaining high-resolution photographs to be used for the selection of landing sites for the forthcoming manned landings. The first three Lunar orbiters, designated LO I, II and III, were placed in equatorial orbits and were so successful in obtaining high-quality images of candidate sites that the two remaining

Figure 2.4 High-resolution television picture returned by Ranger 9 at an altitude of 173 km above the lunar surface 1 minute 12 seconds before impact in the crater Alphonsus. The rim of the crater is on the right; prominent fissures cut the crater floor. Elongate, dark-halo craters on the fissures (arrow) are thought to be volcanic vents. Field of view is 26 by 31 km (Ranger 9 frame B75).

Table 2.2 Milestones in geologic exploration of the Solar System.

Encounter Date		Spacecraft	Event
4 Oct.	1957*	Sputnik 1	First spacecraft
12 Sept.	1959*	Luna 2	Lunar landing (impact)
4 Oct.	1959*	Luna 3	Photograph of lunar farside
26 Aug.	1962	Mariner 2	Venus flyby
1 Nov.	1962*	Mars 1	Mars flyby
28 July	1964	Ranger 7	Lunar "hard" landing
26 Nov.	1964	Mariner 4	Mars flyby
16 Nov.	1965*	Venera 3	Venus landing (impact)
31 Jan.	1966*	Luna 9	Soft landing and pictures
31 Mar.	1966*	Luna 10	Lunar orbiter
30 May	1966	Surveyor 1	Controlled soft landing
10 Aug.	1966	Lunar Orbiter 1	Lunar photographic orbiter
15 Sept.	1968*	Zond 5	First life forms to Moon and back
21 Dec.	1968	Apollo 8	First men to Moon, no landing
16 July	1969	Apollo 11	First manned lunar landing
17 Aug.	1970*	Venera 7	Soft landing on Venus
12 Sept.	1970*	Luna 16	Unmanned sample return
19 May	1971*	Mars 2	Mars orbiter
28 May	1971*	Mars 3	Mars lander
30 May	1971	Mariner 9	Mars orbiter
3 Mar.	1972	Pioneer 10	Jupiter flyby
5 Apr.	1973	Pioneer 11	Saturn flyby
3 Nov.	1973	Mariner 10	Mercury flyby
8 June	1975*	Venera 9	Surface pictures
20 July	1976	Viking 1	Mars soft landing
20 May	1978	Pioneer Venus	Venus orbiter
5 Mar.	1979	Voyager 1	Jupiter flyby
13 Nov.	1980	Voyager 1	Saturn flyby
June	1985*	Vega 1, 2	Weather balloons, Venus
Oct.	1983*	Venera 15/16	Venus radar mapping
24 Jan.	1986	Voyager 2	Uranus flyby
6 Mar.	1986*	Vega 1	Comet Halley flyby
24 Aug.	1989	Voyager 2	Neptune flyby
10 Aug.	1990	Magellan	Venus radar orbiter
29 Oct.	1991	Galileo	Asteroid Gaspra flyby
28 Aug.	1993	Galileo	Asteroid Ida flyby

*Soviet missions.

orbiters were devoted toward scientific objectives. Both were placed in polar orbits (enabling any part of the Moon to be observed); Lunar Orbiter IV was placed in a relatively high orbit and obtained images that were used for making global maps; Lunar Orbiter V was placed in a lower orbit and was commanded to obtain high-resolution images of geologically interesting sites.

Lunar orbiter, nick-named the "flying drugstore" because of its ability to develop its film in flight, returned a wealth of photographs (Fig. 2.5) and documented in exquisite detail not only features which had been so tantalizing from Earth-based views, but also an array of totally unexpected features. Nearly 99 percent of the Moon was photographed by the Lunar orbiters and these images are still an important resource for lunar geologic studies.

Surveyor spacecraft were "soft" landers, designed to function after touchdown. Although the primary purpose of the Surveyors was to obtain pictures of the surface (Fig. 2.6) and was aimed toward gathering information on the physical properties of the lunar surface, later spacecraft in the series were equipped with instruments for determining compositions of the lunar surface. An alpha particle scattering device on board Surveyors V and VI showed that the composition of the lunar maria was similar to that of basalt, lending support to interpretation that mare areas were of volcanic origin.

The Apollo missions to the Moon mark a technological and scientific achievement unparalleled in history, and hundreds of accounts of this milestone in exploration have been written. The Apollo series included manned orbital excursions to the Moon as precursors to the landing. During these pre-landing missions, photographs from Apollos 8 and 10 played a key role in documenting not only the engineering aspects of the flight, but also in providing for the first time high-resolution color images of the Moon and the Earth from deep space.

Although the first three manned landings, Apollos 11 (Fig. 1.3), 12 and 14, were primarily engineering missions, their returned samples from the Moon were intensively analyzed in laboratories on Earth. Apollos 15, 16 and 17 carried much more extensive scientific payloads, both for the landers (including the lunar "rovers" which gave the astronauts greater mobility on the surface) and for the command module (CM) which remained in lunar orbit during the lander phase. While in orbit, the CM obtained high-quality photographs and utilized an array of other remote-sensing instruments to obtain geochemical and geophysical data (Fig. 1.5).

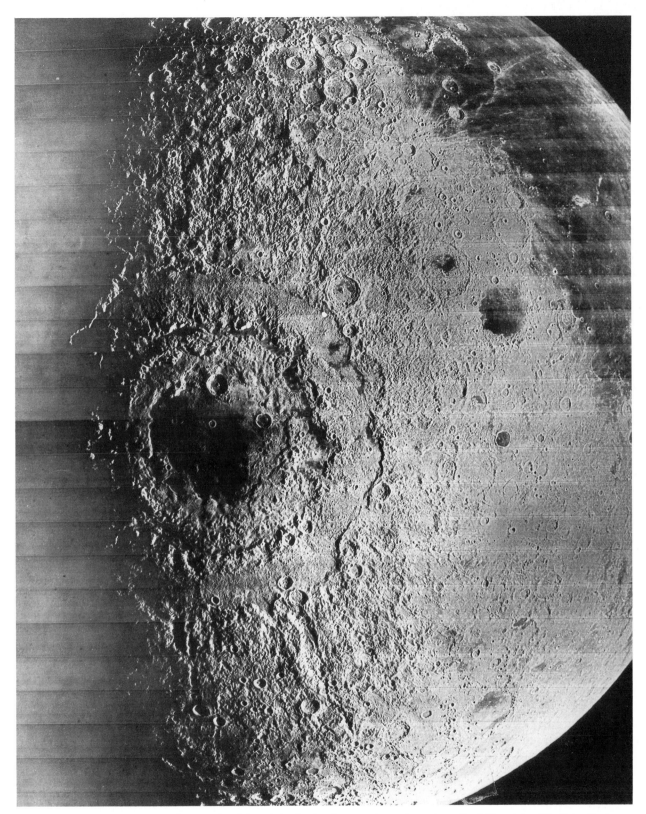

Figure 2.5 Lunar orbiter image of the 900 km Orientale basin showing prominent rings—Inner Rook Mountains (A); Outer Rook Mountains (B); and Cordillera Mountains (C)—and the central filling by mare lavas. Only the far right (eastern) side of this basin can be seen from Earth. Note the radial pattern formed by ejecta from the impact basin. The outer two mountains rise more than 3 km above the adjacent terrain and are some of the highest features on the Moon. Spudis (1982) estimates that the transient excavation cavity was about 570 km across. Lunar orbiter images can be recognized from the *framelets* (strips) which have been mosaicked in the image restoration process after transmission from the spacecraft (Lunar Orbiter IV-M187).

Figure 2.6 Mosaic of Surveyor 7 images taken on the surface of the Moon in January 1968. Surveyor images were the first to show the lunar surface in high resolution. Surveyor 7 landed on the rim of the crater Tycho (see Fig. 2.7). Largest rock in view is about 60 cm across; it is probably a block of ejecta from Tycho and was responsible for the small crater to its right (NASA photograph 83 H 40).

Concurrent with the US manned lunar program, the Soviets carried out a series of successful unmanned lunar missions which included automated roving spacecraft— the Lunakhod—and returned samples of the Moon. Some of these spacecraft were equipped with drills which enabled cores to be taken from the subsurface. Figure 2.7 shows all lunar landings, manned and automated, from US and USSR missions.

The combined results of the US and USSR missions to the Moon provided a remarkable wealth of data during the early 1970s. Never before had so many scientists and laboratories been involved in such a focused study. Most of the results of the lunar program can be found in the annual Proceedings of the Lunar and Planetary Science Conference (a meeting held annually at the Johnson Space Center in Houston), published as a supplement to *Geochimica et Cosmochimica Acta* by Pergamon Press (1970–81), a supplement to the *Journal of Geophysical Research* (1982–87), by Cambridge University Press (1988–9), and by the Lunar and Planetary Institute from 1990.

Concurrent with the lunar program, both the US and the USSR began exploration of planets beyond the Moon. Partly by chance, partly by design, and partly as a result of various mission failures, the US tended to focus on Mars while the Soviets concentrated on Venus, at least in the early stages of exploration. **Mariner** and **Pioneer** constitute two families of spacecraft that were the main US "buses" to carry instruments throughout the Solar System. Mariner spacecraft were 3-axis stabilized probes which carried high-resolution imaging systems as well as other instruments; Pioneer probes were spin-stabilized spacecraft which were used primarily for non-imaging experiments.

Between 1964 and 1971, Mariners 4, 6, 7 and 9 successfully flew to Mars and returned more than 8000 im-

Figure 2.7 Map of the lunar nearside showing various landing sites; includes five Surveyor(s) landings (S), six Apollo sites (A), two Lunakhod roving vehicles (LK), five Luna missions (L), of which three (Lunas 16, 20, 24) returned samples to the Earth.

ages of the surface, some with resolutions better than 100 m. In 1976, the Viking mission marked the first successful landing on Mars. Consisting of two Mariner-class orbiters and two "soft" landers, the Viking mission returned nearly 60,000 pictures from orbit (Fig. 1.1), 4000 pictures from the two landers and vast quantities of data from other experiments. In 1988 two Soviet Phobos spacecraft were launched to Mars; contact with Phobos 1 was lost shortly after launch and contact with Phobos 2 was lost after return of limited data from orbit around Mars. Following this string of bad luck, contact with the US Mars Observer spacecraft was lost just as it was to go into orbit in August, 1993.

The Soviet efforts to Venus began with the first successful landing in 1970 by Venera 7. In the ensuing years, landings have been accomplished by Veneras 8, 9, 10, 11, 12, 13 and 14 as well as the Vega 1 and 2 missions, and the Soviets have been able to obtain images of the surface (Fig. 2.8), determine chemical compositions of surface materials and measure surface winds and temperatures. These are remarkable achievements given the extremely hostile environment on the venusian surface; tem-

Figure 2.8 Venera 13 image of the surface of Venus (courtesy NASA-Jet Propulsion Laboratory).

23

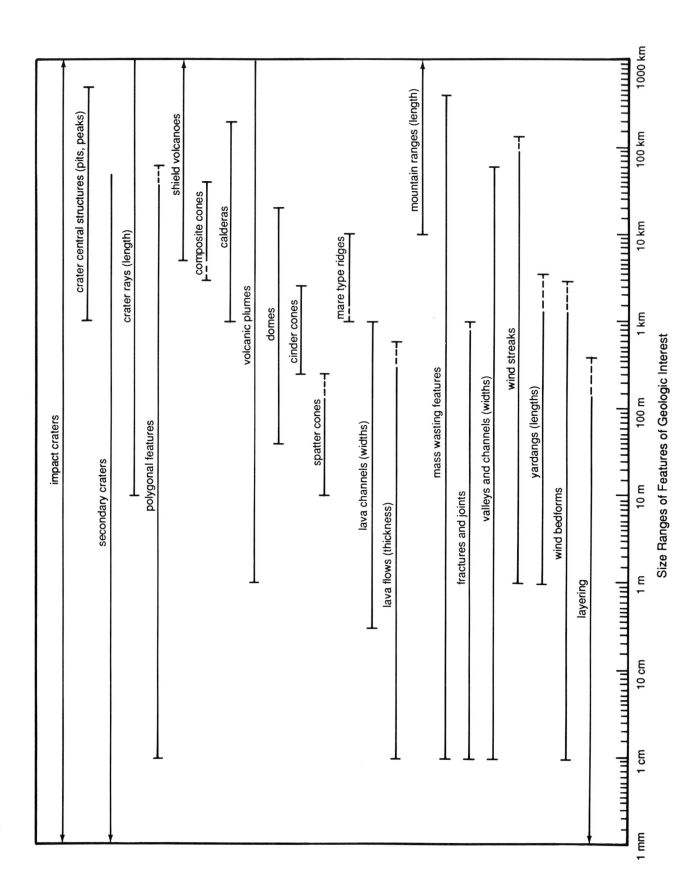

Size Ranges of Features of Geologic Interest

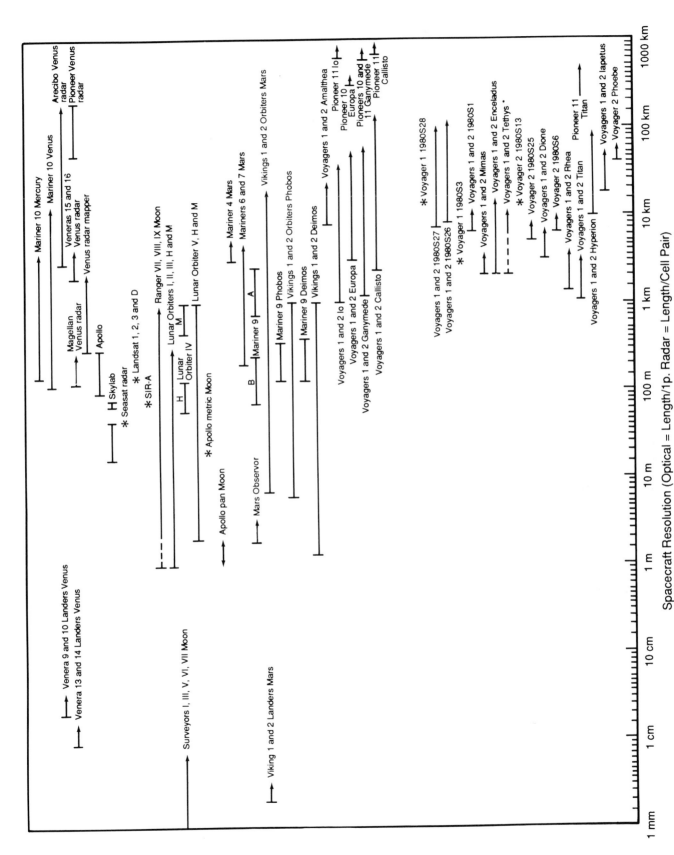

Table 2.3 Resolution required to detect various surface features compared with ranges of resolution from lunar and planetary missions.

Spacecraft Resolution (Optical = Length/1p. Radar = Length/Cell Pair) peratures average 753 K, atmospheric surface pressure is about 90 bar, and the atmosphere is laced with sulfuric acid. In the fall of 1983, Veneras 15 and 16 were placed in orbit and began returning moderate resolution (~1–3 km) radar images of the northern hemisphere of Venus.

The US Pioneer Venus mission was important because it provided our first maps of Venus. Since dense clouds completely hide the surface from imaging by conventional camera systems, radar imaging is the only means for "seeing" the surface of Venus from above the atmosphere. Although not designed to obtain images, the radar system on board Pioneer Venus was adapted to acquire topographic data. In the geologic context, the US Magellan mission provided the first high-resolution global images showing the diversity of the surface.

Mercury has received relatively little attention in Solar System exploration. In 1973, Mariner 10 made the first reconnaissance flyby and obtained high-resolution images. From serendipitous orbital geometry and clever manipulations by the engineering teams at the Jet Propulsion Laboratory, Mariner 10 was able to loop around the Sun and make not only a second flyby of Mercury, but a third pass as well, acquiring images and other data each time. In all, slightly less than half of the planet was observed by Mariner 10, the only mission to Mercury.

Thus far, only the US has reached beyond Mars to explore the outer Solar System. In 1972, Pioneer 10 made the first flyby of Jupiter and was followed 12 months later by Pioneer 11. In 1979, Pioneer 11 reached Saturn; both Pioneers obtained images during the planetary encounters. Although crude, these images showed variations in surface markings for some of the Galilean satellites and piqued scientific curiosity.

Voyager is by far the most successful mission in the reconnaissance of the Solar System. Consisting of two Mariner-type spacecraft, Voyager 1 and Voyager 2, this mission began with flybys of Jupiter in 1979 and Saturn in 1980 and 1981 and continued with flybys of Uranus

and Neptune in 1986 and 1989 respectively. In all, the Voyager spacecraft returned more than 100,000 images of the outer planets (except Pluto), their ring systems, and most of their satellites, including many that were discovered on the images.

Solar System exploration has made remarkable advances since the first simple probes were sent to the Moon. Although the depth of knowledge is highly variable, sufficient data have been gathered from these missions to describe the major physiographic provinces of about 80 percent of the planets and their satellites.

2.4 Planetary images

The missions described in the last section carried a wide variety of remote sensing systems on their journeys of exploration. It is not the intent to discuss here all the methods and instruments employed in remote sensing—Colwell (1983) provides such a discussion—but rather to review systems important in planetary geology. Basically, most imaging systems consist of three elements: an optical system (i.e., a lens) to focus the image, a shutter to expose the image and a sensor. Sensors are generally of two types, film systems or various electronic devices such as television tubes. Davies and Murray (1971) provide an excellent discussion of the various types of imaging devices used in early Solar System exploration.

Among the many important considerations in designing imaging systems, **spatial resolution** (how small an object can be seen) and **spectral sensitivity** (how much energy from a given part of the electromagnetic spectrum is required to see an object) are the most important. Detail is progressively lost as resolution is degraded. Table 2.3 shows resolutions required to detect various geomorphic features on planets.

Figure 2.10 Panoramic camera view from Apollo 16 showing horizon to terminator view over typical lunar highlands terrain; the swirling pattern to the south (lower right) of crater Al-Biruni remains enigmatic (Apollo 16 AS 16-5533).

Figure 2.9 Lunar orbiter film scanning system (courtesy NASA).

2.4.1 Film systems

By far the highest resolution images are produced on photographic film. To utilize the high-resolution capability of film, however, the film must be returned to Earth. Thus far in planetary exploration, this has been achieved only from Soviet Zond and the manned Apollo missions to the Moon. Of the unmanned US lunar and planetary missions, only Lunar orbiter utilized film sensors, but the film was not returned to Earth; it was developed on board and then the image was transferred to Earth as an electronic signal.

Lunar orbiter systems consisted of two cameras which used a common supply of 70 mm film; one camera had a telephoto lens (designated *H* for high resolution) and the other camera had a wide angle lens (designated *M* for medium resolution). Film processing was accomplished with the Kodak Bimat diffusion technique in which a gelatin developer was laminated with the film. After processing, the negative film was passed through a system that converted the images to electrical signals by scanning the film with a microscopic spot of high intensity light, as shown in Figure 2.9. The intensity of the light reaching a photomultiplier tube was modulated by the density (gray tone) of the image on the film.

The electronic signal from the photomultiplier tube was transmitted to Earth where the signal was used to illuminate a glow tube which exposed fresh film, thus generating a duplicate of the image on the spacecraft. Strips of 35 mm film were used in this process to correspond to the scanning sequence of the original image. Consequently, Lunar orbiter photographs are characterized as being composed of strips of images (Fig. 2.5). Although some stereoscopic views of the lunar surface were made from the Lunar orbiters, the continuity between strips was lost in the electronic transformations, resulting in a stair-stepping effect when viewed stereoscopically.

The Apollo missions to the Moon all involved camera systems using film which was returned to Earth. Returned film enabled analysis without the complexities of electronic transformation and provided reliable stereoscopic models useful for photogrammetry. Most simple were the hand-held cameras using 70 mm film. These were used both from orbit and on the ground (often used to make panoramic views from overlapping frames which were mosaicked) and involved both color and black and white film.

Although a sophisticated mapping camera was carried on the Apollo 14 mission, a malfunction early in the mission resulted in very little usable data. Apollos 15, 16 and 17 all carried two high-quality camera systems, a panoramic camera for detailed geologic studies, and a metric (mapping) camera which enabled topographic data to be derived. The "pan" camera, using a 610 mm focal length lens, was highly sophisticated and obtained stereoscopic views of the surface with resolutions as good as 1 m from an orbital altitude of 100 km. The camera rotated continuously in a direction across the path of the orbiting spacecraft, giving a panoramic view (Fig. 2.10).

The metric camera was a mapping system which consisted of two cameras, one pointing downward and one pointing away from the Moon. Both used 76 mm lenses,

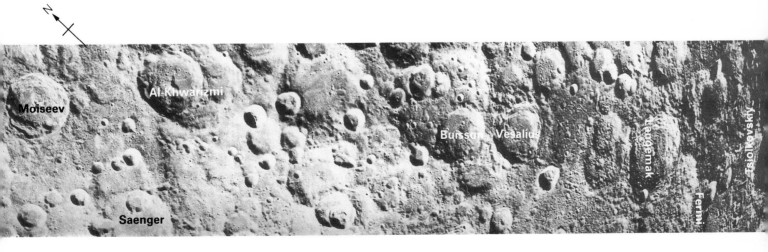

with the upward-pointing camera being used to fix precisely the geometric position of the spacecraft in relation to the star field, with the Moon's surface being photographed by the other camera. Although the spatial resolution of metric frames is only about 20 m, the precision of location enabled high-quality maps to be produced.

2.4.2 Television systems

The imaging system most widely used in planetary exploration in the 1970s and 1980s involved television, or vidicon, cameras. Vidicon systems employ a small electron gun and a photoconductor. The image is optically focused on the photoconductor so that a beam of electrons from the gun is transformed into a current which varies with the intensity of the light reflected from the scene. This current can be in either analog (flown on early systems, such as Ranger) or digital format and can be stored either on tape on board the spacecraft, or transmitted directly to Earth.

Vidicon images from the Mariner and Voyager missions were compiled as arrays of picture elements (pixels), arranged in lines. The spatial resolution of a vidicon image is determined by the distance of the spacecraft from the object, the characteristics of the lens system and the size of the individual pixels.

2.4.3 Facsimile systems

Facsimile systems are used every day in countless newspaper offices around the world to transmit pictures over telephone lines. Basically, a scene is divided into a grid of pixels and each pixel is assigned a gray level–essentially the same idea as for vidicon pictures. However, unlike the vidicon system in which the whole image is focused onto the tube at one time, facsimile cameras scan across the scene by rotating a slit across the scene as a panorama, shown in Figure 2.11. Facsimile cameras were used on the two martian Viking lander spacecraft (Fig. 1.4) and on the Venera landers (Fig. 2.8).

Figure 2.11 Diagram of facsimile camera and related system for the Viking Lander camera flown to Mars (courtesy NASA).

2.4.4 Radar-imaging systems

Dense clouds completely obscure two objects of geologic interest—Venus and Titan (one of the moons of Saturn)—and hide their surfaces from imaging by any of the optical systems described above. Using long wavelength energy capable of penetrating clouds, radar-imaging systems were developed to explore Venus and possibly Titan from spacecraft.

Radar is considered an "active" form of remote sensing because the imaging system generates energy as a radar beam which is then reflected from the scene and received by the sensor. Thus, in effect, radar-imaging systems provide their own "light" and can operate regardless of time of day.

The Soviet Venera 15/16 and the US Magellan missions carried radar imaging systems in orbit around Venus to obtain the first detailed views of the surface. Seasat and the Shuttle Imaging Radar (SIR) were radar-imaging experiments flown in orbit around Earth which provided important insight into the use of radar for geologic interpretation (Fig. 2.12).

Figure 2.12 Radar image from the SIR-A (Shuttle Imaging Radar) system flown on the Columbia spacecraft showing a 50 km wide strip from the Pacific shoreline of Chile (left side) across the western Andes and the Altiplano of Bolivia. Features visible include stream patterns (A), dome volcano with summit crater (B), stratovolcano (C), and wind streaks (D) formed on smooth sedimentary plains (courtesy NASA, Jet Propulsion Laboratory).

2.4.5 Charge-coupled devices

Charge-coupled devices (CCD) were invented in 1969 by Bell Laboratories and are used in a variety of communication systems and as a solid-state imaging device. CCD "chips" consist of one layer of metallic electrodes and one layer of silicon crystals, separated by an insulating layer of silicon dioxide. When used as an imaging system, the CCD chip is structured as an array of elements; light focused on the chip by a lens causes a pattern of electrical charges to be created. The charges on each element are proportional to the amount of light and provide an accurate representation of the scene. Each charge can be transmitted separately and then reconstructed as pixels using conventional image-processing techniques.

2.5 Digital image processing

Vidicon and facsimile imaging systems used by Mariner, Viking and Voyager missions and the CCD systems to be used by the Space Telescope and flown on Galileo all include digital signal processing for transmission to Earth. Because each pixel is in digital format, the images can be manipulated by computers using various image-processing techniques. Image processing has become an extremely important aspect of planetary exploration, and knowledge of the fundamentals is critically important for geologic utilization of spacecraft digital images. Although many books have been written on the subject, an excellent brief introduction to the fundamentals of digital image processing is given by Condit and Chavez (1979).

A digital picture is a two-dimensional array of pixels, each of which is assigned a value representing the light brightness (called **DN** or **digital number**). The accuracy of the brightness level is a function of the digital encodement; for example, 8-bit (2^8) encodement allows 256 levels of gray to be assigned to each pixel (0 = black, or below the threshold of sensitivity; 255 = white, or saturation). This range of gray levels far exceeds the dynamic range of any photographic film, which means that digital images contain much more information than can be recorded in camera systems using film.

Horizontal rows of pixels are called **lines** and vertical columns are called **samples**. The maximum pixel format used in any planetary mission flown thus far was the Viking orbiter camera system involving 1056 lines by 1204 samples, with each pixel transmitted in 8-bit encodement. Thus, each complete image consists of 10.2 million computer bits of data.

Computer programs for image processing can be classified into three types for geologic analysis: (a) image correction programs, (b) image enhancement programs, and (c) multispectral utilization programs.

Figure 2.13 Mariner 9 image showing "zipper" pattern (A; camera "noise") and white streaks (B) generated from the reseau marks; both of these artifacts can be removed by image processing. The dark spot (C) results from a dust speck in the camera system and is too large to be "removed" by processing (Mariner 9 4233-31).

2.5.1 Image correction programs

Image correction programs are used to remove electronic noise, replace lost data, correct for variations in sun angle, and make calibration corrections for the camera system. For example, during electronic transmission of the image data, "noise" may be introduced which superimposes a pattern on the image, as shown in Figure 2.13, or data for individual pixels, or lines of pixels, may be lost. In order to generate a "whole" picture, DN values are assigned to the "noisy" or "lost" pixels by determining the average value from the surrounding pixels and the image is then filled in using these values.

2.5.2 Geometric corrections

Geometric distortions in images also can be corrected through image-processing techniques. For example, Figure 2.14 shows an oblique view of part of Mercury. Through knowledge of the camera geometry and location of the spacecraft, individual pixels can be shifted in their position on the image to correspond to a vertical view in which the scale of the image is orthogonal (Fig. 2.15(a)). Such geometric transformations are important for analyses of sizes and shapes of landforms, and for mosaicking sequences of pictures. Geometric transformations can be made for almost any cartographic format, including Mercator, Lambert conformal or polar stereographic projections.

2.5.3 Enhancement techniques

Enhancement programs are used to emphasize different aspects of the scene on digital images. Although numer-

Figure 2.14 Specially enhanced Mariner 10 image of Mercury (courtesy M. Davies; Mariner 10 FDS 27321).

ous programs have been developed for image enhancement, the most common programs involve **stretching** and **spatial filtering**.

Figures 2.16 and 2.17 show a series of image-processing routines involving both image-correction and image-enhancement techniques. The surfaces of some planets are relatively low in contrast and "stretching" of the DN values can enhance the contrast on the image to bring out the surface detail, as shown in Figures 2.16(a) and (b) for a typical scene on Mars. In this view, most of the levels of gray are within the same narrow range, shown by the DN histogram. These gray levels can be "stretched" over a wider range to produce the view shown in Figure 2.16(c). The "stretch" can be accomplished using any one of a number of algorithms, including linear stretches (an even spread of the histogram over a given range of DN values), or in various gaussian distributions.

Spatial filtering changes pixel DN values according to the values of neighboring pixels, either as averages, or as differences. **High-pass filters** amplify rapid changes from one pixel to another and are used to sharpen edges and enhance small detail, although there is generally a loss of tonal variations (Fig. 2.17(a)). **Low-pass filters** enhance low-frequency detail and tend to smooth image detail, while preserving tonal quality such as albedo (Fig. 2.17(b)).

2.5.4 Multispectral images

Various color pictures are generated from multiple images of the same terrain taken through colored filters as part of most spacecraft camera systems. Each selected range of wavelength (determined by the appropriate filter) is

Figure 2.15(a) Same frame as shown in Figure 2.14 which has been corrected for camera shading and variation in scene luminance due to sun angle and transformed to a Mercator projection; processing was also done to enhance albedo markings (courtesy M. Davies).

Figure 2.15(b) Shaded airbrush chart of the area shown in (a) of part of the Kuiper quadrangle, Mercury (from Davies & Batson 1975).

663A38 RAW IMAGE

Figure 2.16(a) "Raw" or unprocessed Viking orbiter image containing dropped line of data across the picture and reseau marks (evenly spaced black dots used for geometric analyses). Distribution of DN values shown in histogram (Arizona State University Image Processing Facility).

0663A38 RESEAUS, DROPPED LINES, AND ZIP REMOVED

Figure 2.16(b) Image in which correction routines have been applied to remove the reseaus and the dropped line by filling in pixels based on averaging values of surrounding pixels (Arizona State University Image Processing Facility).

Figure 2.16(c) Same image which has been stretched to enhance contrast (as indicated by the histogram), in addition to being corrected and calibrated (Arizona State University Image Processing Facility).

registered on separate images. These multiple images are then registered geometrically to generate a color view. Except for color film returned from the Moon, all color images of the planets have been obtained in this manner and are termed **false color** images.

By ratioing the DN values of images taken through different filters of the same scene, the spectral reflectances, or color difference between rock units, are often enhanced. Thus, color ratioing is often a useful technique for photogeologic mapping.

2.5.5 *Other image-processing techniques*

The development of image-processing techniques is an ever-expanding field as more researchers become familiar with the general process and as new applications are discovered. Some of the more important techniques for planetary geology include **computer mosaicking** (Fig. 2.18), **picture differencing** in which images taken at different times of the same area are subtracted from one another to detect possible changes (Fig. 2.19), and combining multiple data sets to determine possible correla-

tions (such as merging multispectral images with radar or infra-red images).

Because of the manner in which spacecraft images are acquired and processed, artifacts are frequently introduced. In the decade and a half of planetary digital image processing, techniques have been devised to reduce these artifacts; nonetheless, a great many problems exist in using digital images. Furthermore, use of images from earlier missions requires familiarity with these artifacts. Some of the more common artifacts and problems are illustrated in Figure 2.20.

Despite the difficulties encountered in digitally processed images, the benefits derived from image-processing techniques far outweigh the disadvantages and will continue to be an important part of planetary exploration.

2.6 Planetary cartography

Maps are an essential part of planetary exploration. They provide the base for plotting observations and give a context for the drawing of conclusions.

33

0663A38 HIGH PASS FILTER 31 X 31

Figure 2.17(a) Same image as in Figure 2.16, but a *high-pass filter* has been applied to sharpen edge features such as crater walls (Arizona State University Image Processing Facility).

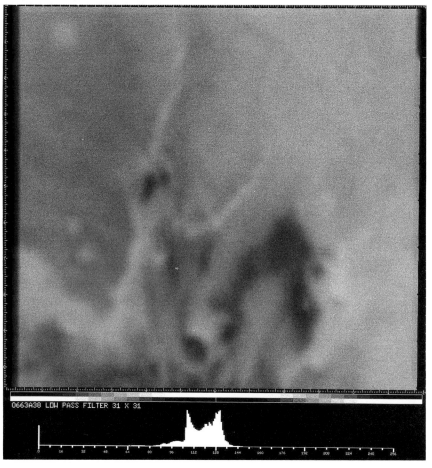

0663A38 LOW PASS FILTER 31 X 31

Figure 2.17(b) Same image in which a low-pass filter has been applied to smooth image detail while preserving tonal quality.

Figure 2.18(a) Mosaic of Mariner 10 frames assembled by hand, showing the Caloris basin of Mercury.

Figure 2.18(b) Same area as (a), but computer-generated mosaic in which frame boundaries have been "processed-out" and the entire scene is filtered and enhanced as a single image (NASA images courtesy of Jet Propulsion Laboratory).

Beginning with Galileo's simple sketch maps of the Moon, planetary cartography has advanced to a sophisticated science, as outlined by Greeley and Batson (1990). The typical sequence of generating planetary maps begins with the development of an uncontrolled mosaic of images covering a general region. Control points, such as well-defined and identifiable surface features, are established within the region and are used to generate a controlled reference network, or planetary grid system. The second stage is to produce a controlled mosaic of images which geometrically fits the network. This mosaic is generated in some conventional format, such as a Mercator projection, and may have additional information superimposed, such as contour lines, to produce a photo-topographic map. Figures 2.21 and 2.22 show the layout of map quadrangles for the major planets and satellites.

Because the images composing the controlled mosaic may have been taken under a wide range of lighting conditions and may be of different spatial resolutions, it is often desirable to produce a more uniform cartographic product. The **shaded airbrush relief map** serves this purpose. Using available images for detail and the controlled mosaic as a base, skilled artists generate these renditions which then serve as uniform base maps. Shaded airbrush maps may have additional information superimposed, such as contour lines, albedo and place names, as discussed by Inge and Bridges (1976) and shown in Figure 2.23.

More than 1600 maps of the planets and satellites have been produced by the US Geological Survey and the Defense Mapping Agency. Inge and Batson (1992) provide an excellent index and catalog to these maps, organized by planetary object.

In the early 1970s, when it became evident that topographic nomenclature would soon be required for a large number of planetary bodies in the Solar System, the International Astronomical Union organized its Working Group for Planetary System Nomenclature. Modeled on

a

b

c

Figure 2.19 Image processing for change detection ("picture difference"). (a, top left) Mariner 9 image (DAS 08189414) showing dark patches inside a large crater, (b, left) same crater imaged later in the mission (Mariner 9 image DAS 10649609), (c, above) image of the "difference" between (a) and (b) in which the DN values of (a) were subtracted from (b) and the resultant DN values used to construct a new image. The dark areas in (c) represent newly darkened zones between the time (a) and (b) were observed.

earlier working groups for lunar and martian nomenclature, this new Working Group, divided into five task groups (Moon, Mercury, Venus, Mars and Outer Solar System), provided a formal system for naming surface features. The names used are applied to the features according to systematic plans and are drawn from many sources, including famous deceased scientists and artists, gods and goddesses from mythology, classical places and even small modern towns of Earth.

2.7 Summary

Nearly two decades of Solar System exploration have resulted in a wealth of planetary data. Most geologic information for the Moon has stemmed from the US unmanned missions of the Ranger series, the Surveyor program, the Lunar orbiters and the US manned Apollo program. Soviet contributions have come from the Zond and Luna missions. Geologic exploration of planets beyond the Moon has been accomplished by the US Mariner 10 mission to Mercury (with Venus flyby), the Soviet Venera series of orbiters and landings on Venus, the US Pioneer Venus orbiter/probe and Magellan radar orbiter, the US Mariner and Viking missions to Mars, the Soviet Mars missions, and the US Pioneer, Voyager and Galileo missions to the outer Solar System.

An integral part of nearly all these missions has been various imaging experiments. Planetary images provide the primary data base to determine the physiography of the planets and to interpret the processes that have shaped their surfaces.

Figure 2.20 Examples of various artifacts and blemishes in spacecraft images (see also Fig. 2.13): (left, top and center) "doughnuts" (arrows) resulting from dust specks in the Viking orbiter camera system; these artifacts could be mistaken for surface patterns on Mars, (top image, VO 826A68; center image, VO 846A40); (bottom left) an oblique Viking orbiter image showing "blocky" pixels which result from problems in transmission of data from the spacecraft to Earth (VO 815A63); (top right) "bimat bubbles" (arrow) in Lunar orbiter photograph of the Moon result from tiny air pockets between the film layer and developer layer (part of LO IV-67H₁), (center right) "wormy" texture in Apollo metric photographs that results from processing shows in both photographic prints and lunar photomosaics, and (bottom right) tiny white crosses are not blemishes in Lunar orbiter photographs but are reseau marks built into the camera system; some non-scientists have interpreted these as representing cultural features by intelligent life on the Moon! (Lunar Orbiter V-H67).

37

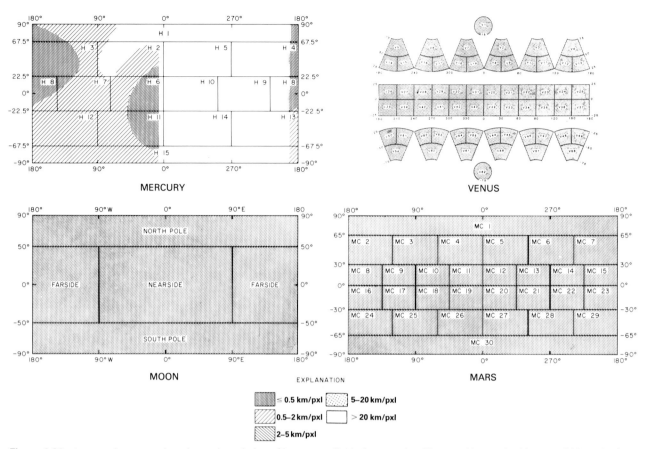

Figure 2.21 Layout of map quadrangles and resolution of images available for mapping Mercury, Venus, the Moon and Mars. As the term is used here, resolution is the size of a picture element of the surface of the planet (from Batson 1981).

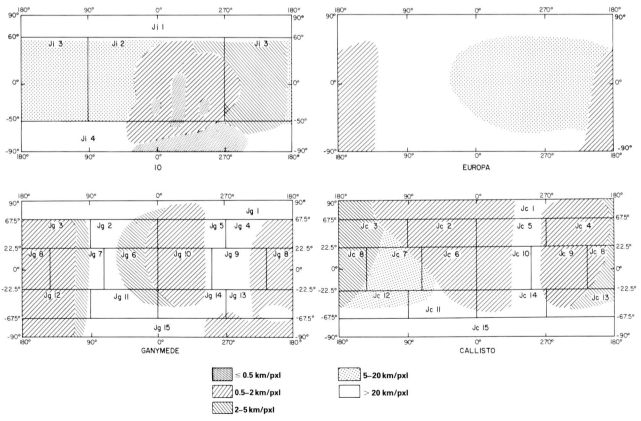

Figure 2.22 Layout of map quadrangles and resolution of images available for mapping Io, Europa, Ganymede and Callisto (from Batson 1981).

38

Figure 2.23 Mariner 9 image (a, left) of part of Mars and corresponding airbrush drawing (b, right) which reverses the illumination, removes camera artifacts and incorporates data from other images (courtesy P. Bridges, US Geological Survey, Flagstaff).

3 Planetary morphologic processes

3.1 Introduction

Geomorphologists have long recognized that the surface of the Earth is shaped by various processes. These processes are generally placed into three primary groups: (a) **tectonic processes**, involving deformation of the lithosphere; (b) **volcanic processes**, involving eruptions of magma onto the surface; and (c) **gradational processes**, involving the erosion, transportation and deposition of surface materials by various agents, such as wind and water, to bring surfaces to a more uniform level. In the planetary context, we must add a fourth major process, **impact cratering**, which involves collision of solid objects.

Each process produces landforms, some of which are diagnostic of the processes involved in their formation. Learning to recognize these landforms is one of the primary goals of planetary geology, for it is through this recognition and the interpretation of the relevant processes that geologic histories can be derived for the planets and satellites.

In this chapter, each of the major surface-modifying processes is described and illustrated. Because most of the knowledge about these processes has been obtained from field and laboratory studies on Earth and provides the "ground truth" for extraterrestrial interpretations, most of the examples shown will be taken from Earth. Some processes, however, are understood far better on other planets and, in those cases, extraterrestrial examples will be used.

It should be noted that not all geomorphologic processes will be reviewed here; rather, only those processes that seem to be appropriate for planetary geology will be considered. The potential flaw in this selectivity is that some processes which are critical for planetary interpretations may be neglected. However, to attempt to include a complete review of geomorphology is not the intent, and the reader is referred to the numerous excellent textbooks on the geomorphology of the Earth, including King (1967), Twidale (1976), Bloom (1978) and others.

Included with the description of each process is a brief discussion of how the different planetary environments might affect the process and the resulting landforms.

3.2 Impact cratering

The American geomorphologist William Morris Davis (1926; reviewed by Roddy (1977)), summarized the controversy surrounding the origin of lunar craters with the following statement:

> It has been remarked that the majority of astronomers explain the craters of the moon by volcanic eruption—that is, by an essentially geological process—while a considerable number of geologists are inclined to explain them by the impact of bodies falling upon the moon—that is, by an essentially astronomical process. This suggests that each group of scientists find the craters so difficult to explain by processes with which they are professionally familiar that they have recourse to a process belonging in another field than their own, with which they are probably imperfectly acquainted, and with which they therefore feel freer to take liberties.

Planetary research has resolved many of the misconceptions regarding impact craters and has laid the foundation for understanding this important geologic process. Several conferences have been held on impact cratering mechanics and related phenomena, with proceedings publications which are important to planetology (French and Short 1968, Roddy *et al.* 1977, Silver and Schultz 1982).

Impact cratering involves the nearly instantaneous transfer of energy from an impacting object, called the **bolide**, to the target surface. Bolides can include meteoroids, asteroids and comets. Velocities of these objects upon impact on Earth range from 5 to 40 km s^{-1}. Using the simple expression for kinetic energy of KE = 0.5 (mv^2), where m is the mass of the bolide and v is the velocity, it can be seen that an average nickel–iron meteoroid 30 m in diameter traveling at 15 km s^{-1} could impart about 1.7×10^{16} J of energy onto a planetary surface, the equivalent of exploding about 4 million tons of TNT! Such an impact event is considered to be responsible for the formation of Meteor Crater in northern Arizona, in which more than 175 million tons of rock were excavated to leave a crater more than 1 km across and 200 m deep (Fig. 1.8(e)). Unlike most geologic processes, an individual impact cratering event is of very short duration; Meteor Crater probably formed in about one minute.

3.2.1 Impact cratering mechanics

No natural impact craters have been observed in active formation. This, together with the very short duration of a cratering event, has led most investigators into the laboratory to conduct cratering experiments where the process could be studied under controlled conditions (Fig. 1.8). From analyses of high-speed motion pictures and detailed analyses of cratered targets, Don Gault and his colleagues (1968) at the NASA Ames Research Center derived the general sequence of cratering (Fig. 3.1). First is the *compression stage*, in which the projectile contacts the target, penetrates the surface and is engulfed, resulting in high-speed jetting of material outward from the zone of contact. At the same time, intense shock waves are sent through both the target and the projectile. In this stage of the impact, shock pressures of several megabars are common, exceeding by three to four orders of magnitude the effective material strength of common rocks. It is this high shock pressure which sets impact cratering apart from other geologic processes. The impact process results in intensely crushed and broken target material, some of which is so severely shock-metamorphosed that some of the rock is melted and vaporized.

The second stage involves *excavation* of the crater as shock waves and attendant rarefaction (or decompression) waves set target material into motion. The material excavated from the crater (Fig. 3.2), termed **ejecta**, is distributed radially as a blanket of fragmental debris. **Continuous** ejecta covers the surface as an uninterrupted blanket from the crater rim outward and gives way to the **discontinuous** ejecta, which in turn grades into zones of *secondary craters*, formed by the impact of ejecta blocks and clods.

The third stage includes various *post-cratering modifications* not directly attributable to shock waves. These include slumping of the crater walls, isostatic adjustments to the floor and rim, and erosion and infill of the crater. The third stage may continue over long periods of time until the crater is eventually obliterated by processes of gradation and viscous flow.

3.2.2 Impact craters on Earth

The existence of impact craters on Earth was not readily accepted by the scientific community. Even to advocates of impact cratering, prior to 1930 fewer than ten impact structures were known on Earth. By 1966, the number had only risen to about 33, but after intensive searches and establishment of criteria for the recognition of impact craters (Table 3.1), by the 1980s nearly 200 craters and related structures had been fairly well documented as resulting from impact processes (Classen 1977).

Figure 3.1 Stages in impact cratering: (a, top) initial stage involves contact of bolide with target and high-speed jetting of upper target materials; (b, center) passage of shock waves through target and subsequent generation of rarefaction waves sets target material into motion as ejecta leading to (c, bottom) formation of crater and emplacement of ejecta. Note the inversion of stratigraphy in the overturned "flap" of the crater rim. Post-impact cratering modification stage, not shown, involves slumping of walls and infill of crater (from Gault *et al.* 1968).

Figure 3.2 The lunar crater Euler (27 km in diameter) showing principal features of a typical impact crater (Apollo 17 AS 17-2923).

Compared to the heavily cratered surface of the Moon, 200 craters on Earth would appear to be an anomalously low number. However, impact cratering involves chance collisions of planetary objects. Statistically, the longer a surface is exposed, the greater the likelihood that it will be struck by an extraterrestrial object. Thus, we would expect to find impact craters only on older surfaces. Because most of the Earth's crust has been recycled by crustal tectonics and modified by gradation, most impact craters on Earth have been destroyed or obliterated, in contrast to the Moon where most of the early history is preserved in its ancient crust (Fig. 1.2). Thus, to some degree the number of craters one can see on planetary surfaces provides a clue to the degree of surface evolution that the planet has experienced during its geologic history.

Meteor Crater in northern Arizona is one of the best preserved and most intensely researched impact craters on Earth. Studies by Shoemaker (1963), Roddy (1977)

and others have shown that the crater was formed between 20,000 and 30,000 years ago in flat-lying sedimentary rocks. Detailed field studies and analyses of drill cores show that the rocks were highly deformed; strata in the rim of Meteor Crater were folded back as an overturned flap (Fig. 3.1), and rocks beneath the floor of the crater were severely brecciated to a depth of about 350 m.

Unlike many lunar craters of the same size, Meteor Crater is distinctly polygonal in plan view (Fig. 1.8(e)), demonstrating the importance that structural control can exert on crater form. The strata surrounding Meteor Crater have a distinct joint pattern which at the time of impact evidently controlled the passage of shock waves and the subsequent excavation of the fragmented rocks, leading to the polygonal outline of the crater which was further enlarged by post-excavation modifications.

Gene Shoemaker predicts that an impact of the magnitude to form Meteor Crater would occur every 50,000 to

Table 3.1 Criteria for the recognition of impact craters (modified from Dence 1972).

Criterion	Characteristics	Reliability
Remote sensing		
Plan view	Distinctly circular; may be modified by slumping, tectonic patterns, or erosion	Fair, but can be attributed to other processes
Rim Structure	Inverted stratigraphy	Definitive
Central zone	Floor lower than surrounding plain; may contain central uplift	Fair, but can be attributed to other processes
Geophysical observations		
Gravity anomaly	Generally negative	Supportive, but not conclusive
Magnetic field	Variable; may be distinct anomaly over melt rock	Supportive, but not conclusive
Seismic velocities	Generally lower in brecciated zones	Supportive, but not conclusive
Ground observations		
Presence of meteorites	Rare except in very young craters	Definitive
Shock metamorphism	Features such as high pressure minerals, impact melt, planar shock features and shatter cones	Definitive
Brecciation	Observed in ejecta, rim and floor of craters	May be attributed to other processes

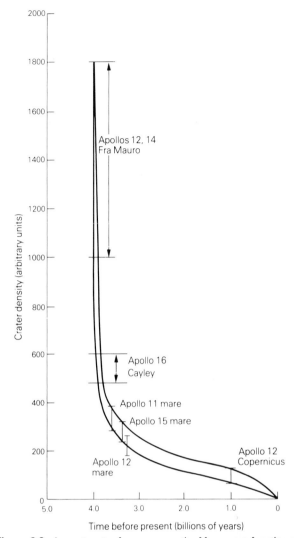

Figure 3.3 Impact crater frequency on the Moon as a function of time, showing that cratering was much more frequent in the early history of the Moon and that there has been a subsequent decay in the rate of impact (after Soderblom *et al.* 1974. Copyright © 1974 Academic Press).

100,000 years on Earth. Although the uncertainties in this estimate are numerous, it gives some appreciation of the frequency of impact crater formation spread over geologic time. Analyses of the crater record on the Moon, which is much better documented than the record for Earth, shows that while smaller impact events are much more frequent, there has been a general decay in the rate of cratering through time (Fig. 3.3).

Field studies of eroded impact craters have yielded important clues about the deformation of the rocks surrounding the crater. For example, the Sierra Madera structure in Texas is estimated to represent a ~16 km in diameter crater that has been deeply eroded. Detailed mapping of the rock units and reconstructions of the original position of the rocks show that the central zone of the structure was uplifted more than 1200 m (Wilshire *et al.* 1972). Such an uplift probably corresponds to the *central uplift*, or *central peak*, of many lunar craters (Fig. 3.2). However, central uplifts are observed only on craters larger than a few kilometers in diameter on Earth (there is no central uplift at Meteor Crater), and it appears that such features require a certain minimum impact energy to form. Although the causes of central uplift are not well defined, they are generally considered to be the result of rapid elastic rebound immediately following the excavation of the crater bowl.

The Ries Kessel of southern Germany has provided insight into the mechanics of ejecta emplacement. This 24 km structure has been the subject of considerable debate among European geologists and traditionally has been attributed to volcanic or crypto-explosion origins. Discovery of the high-pressure minerals *coesite* and *stishovite*, which form only by impact processes, have now convinced most investigators of the impact origin for the Ries Kessel. Extensive surface mapping and drilling to obtain subsurface core samples have provided knowledge on the extent of the ejecta deposit and its properties. The ejecta is composed largely of a brecciated mass known as the Bunte breccia. Petrologic analyses and considerations of ballistics suggest that fragmental ejecta was

Figure 3.4 Mosaic of Viking Orbiter images of the 28 km diameter martian impact crater, Arandes, showing ejecta which is considered to have been emplaced partly as a liquid slurry (from Gault & Greeley 1978).

mixed with melt rock and volatiles as it was excavated from the **transient cavity** (original crater bowl) and thrown in a ballistic trajectory to rain upon the surrounding surface where the material churned up and mixed with local rock. Then the entire mixture continued to slide outward a short distance to settle into place. This model suggests the important role that target volatiles may play in ejecta emplacement, and may be applied in part to the interpretation of certain craters on Mars (Fig. 3.4).

Among the many other structures on Earth that have yielded important data on the morphology of impact craters and the cratering process are the Clearwater Lakes, Canada, which appear to be a double impact crater, the Serra da Canghala structure, Brazil, which displays a central-ring uplift (Fig. 3.5), and the Henbury Craters, Australia, which consist of 13 or more craters, possibly reflecting the breakup of an incoming object to form multiple craters.

3.2.3 Impact crater morphology and effects of different planetary environments

The primary factors governing the size and morphology of impact craters are the impact energy (a function of the size and velocity of the bolide), various properties of the target such as rock strength and the presence or absence of volatiles, and gravity. As discussed by Gault (1974),

gravity affects the cratering process by influencing: (a) the dimensions of the excavation bowl, (b) the extent of the ejecta, and (c) various post-impact crater modifications. For equal-size impact events, fragmented blocks of ejecta could be lifted and excavated more easily on low-gravity planets, leading to larger craters in comparison to high-gravity environments. Furthermore, in low-gravity environments, ejecta is thrown a greater distance, as shown in Figure 3.6. Thus, we would expect to see a wider (but thinner) zone of ejecta surrounding impact craters on the Moon than on higher-gravity planets such as Mars or Mercury. In the modification stages of impact cratering, gravity plays a role by governing the rate of isostatic adjustments, influencing the degree of slumping and perhaps governing the magnitude of potential central uplifts.

Lunar impact craters show a distinctive progression in morphology with increasing size, with simple bowl shapes predominating up to about 10 km in diameter, central-peaked craters occurring at 10–150 km in diameter, clusters of central peaks in the range 100–175 km, peak-ring craters for diameters of 150–220 km, and multi-ringed craters (more commonly called **basins**) for structures greater than about 220 km in diameter. This general progression is observed on other planets as well, but there is a variation in the size ranges and there are additional categories, as will be discussed in subsequent chapters.

The shape of impact craters in map view is controlled by the angle of the incoming projectile and the tectonic

44

Figure 3.5 Oblique aerial photograph of Serra da Canghala in Brazil showing an uplifted central ring and pit, possibly resulting from impact into water-saturated sediments. Outer diameter is 12 km (from Greeley *et al.* 1982; photograph courtesy of John McHone).

Figure 3.6 Comparison of range of heights for ejecta traveling 2 km s^{-1} on the Moon, Mercury, Mars, Venus and the Earth. Note that the high planetary density of Mercury (Table 2.1) permits the ejecta to travel only a little farther on its surface than ejecta on Mars, despite the fact that Mars has a larger diameter than Mercury (from Schultz 1976b).

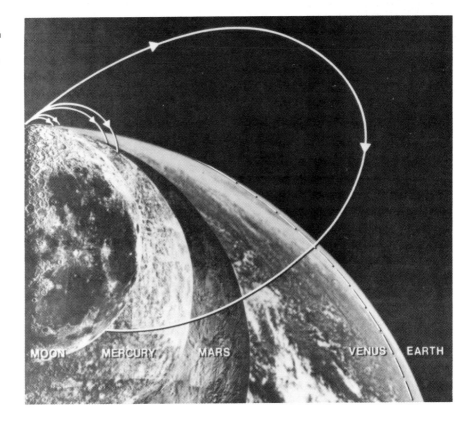

fabric of the target, such as the presence of joints or other fractures, as noted in the case of Meteor Crater. Because impacts involve essentially point-source transfers of energy, both the crater and the distribution of ejecta for most impacts are concentrically symmetrical about the point of impact. Although intuition might suggest that oblique angles of impact would cause elongate craters, experiments have shown that only for very low angles (<15°) do impact craters become noticeably asymmetrical (Gault and Wedekind 1978).

3.2.4 Crater counting as a technique for age determinations

The longer a planetary surface is exposed to impact bombardment, the more craters it should display. Thus, the more craters observed on any given surface (Fig. 1.7), the older it should be, and by counting the number of impact craters on various surfaces, it should be possible to place each surface in a relative age sequence (Fig. 3.7). In principle, if the flux of incoming bolides is known

Figure 3.7 Apollo 15 view of the Moon showing two distinctly different surfaces, the smooth relatively sparsely cratered mare and heavily cratered uplands. Differences in crater frequency can be used for relative age-dating of surfaces. *Sinuous rilles*, the channel-like depressions, are of volcanic origin and were channels for the emplacement of the mare lavas in this area of east central Oceanus Procellarum. Area shown is about 160 km wide (Apollo 15 AS 15-2483).

through geologic time, then it should be possible to assign "absolute" ages for cratered surfaces by dividing the total number of craters observed on a surface by the number of craters formed each year (Hartmann *et al.* 1981).

In practice, however, a great many difficulties arise in using crater statistics for age determinations. Among other problems, these include:

1. *Non-impact* craters, such as those formed by volcanic, karst, thermokarst or other processes, may be indistinguishable from impact craters, and if they are present in significant numbers, the surface would appear anomalously old.
2. *Secondary craters* add to the total crater population and must be taken into account by various models which predict how many secondary craters would form as a function of primary crater size, target material properties, etc. Unfortunately, such models are imperfect, and it is very difficult to determine the presence and number of secondary craters.
3. *Variations in target properties* could cause variations in crater sizes. For example, experiments show that craters formed in targets containing fluids are larger than craters formed in "dry" targets. This difference could cause the size-frequency distribution for the volatile-containing target to be interpreted as representing an older surface.
4. *Crater equilibrium* studies by Gault (1970) have shown that, with time, cratered surfaces reach a stage in which craters of a given size are obliterated by impact erosion at the same rate of formation, as shown in Figure 1.7. Thus, only surfaces that have not yet reached equilibrium for the crater sizes being considered can be analyzed.

Despite these difficulties and uncertainties, impact crater statistics are commonly used as a means for obtaining relative dates of formation for different planetary surfaces. Ages derived from crater counts have been compared with (and calibrated against) radiometric dates obtained from lunar samples and demonstrate the validity of the technique, at least on the Moon where surface-modifying processes are minimal. This result suggests that crater counts may be used to obtain dates for surfaces on other "airless" bodies, such as Mercury. Great caution must be exercised, however, in using crater counts on planets where differences in gradation may occur as a function of latitude, or where significant differences in target properties may alter the crater morphology.

The recognition of impact craters and the understanding of the mechanics of impact cratering as a process remain among the most important areas of investigation in planetary science. As more diverse objects are explored in the Solar System, knowledge of crater form as functions of planetary size and target composition becomes ever more important for the interpretation of surface history.

3.3 Tectonic processes

Deformation of the Earth's crust by tectonic processes is easily demonstrated by features such as faults, folds and fractures, as shown in Figures 3.8 to 3.12. Geologists recognize that these local features can be related, in part, to the style of deformation, such as tension or compression. However, insight into crustal deformation on larger scales was not gained until the unifying concept of global plate tectonics was derived in the 1960s. This insight has been important in understanding the evolution of the crust on Earth and is critical in the interpretation of other planets. More recently, analyses of the styles and timing of tectonism on the terrestrial planets has enabled the derivation of general geophysical models of planetary evolution and thermal history as reviewed by Head and Solomon (1981).

3.3.1 Earth's interior

The interior of the Earth is divided into distinct layers, based on seismologic characteristics and assumptions of composition. The **crust** is a thin zone composed of relatively low-density rocks extending from the surface to a depth of 5–50 km. The boundary between the crust and the underlying **mantle** is marked by an abrupt increase in seismic wave velocities and is called the **Mohorovicic Discontinuity** (the "Moho"), after the Yugoslavian scientist who discovered it. Making up less than 0.1 percent of the total volume of the Earth, the crust occurs in two forms, **sima** and **sial**, terms which are derived from their predominant compositions (Si = silicon, Ma = magnesium, Al = aluminum). Oceanic crust is composed of sima and is relatively thin (~5 km), whereas continental crust is composed mostly of less dense sial overlying a zone of sima and may exceed 50 km in thickness beneath major mountain chains.

The mantle extends to a depth of nearly 3000 km and is composed of materials rich in iron and magnesium. From seismic wave properties, Earth's **core** has been shown to consist of an outer liquid part and an inner solid part, both composed predominantly of iron.

3.3.2 Plate tectonics

In order to understand global tectonic processes, it is necessary to take a slightly different view of the subdivision of the outer few hundred kilometers of the Earth. The crust–mantle–core scheme described above is based principally on compositional differences. Analyses of the physical properties of these zones led to the recognition of the **lithosphere** and **asthenosphere**. The lithosphere consists of the crust and the upper part of the mantle, which together behave as a relatively rigid, solid shell. The lithosphere rests on, and is subject to movements by, the underlying semi-molten and mechanically plastic layer of the upper mantle, termed the asthenosphere.

Within the mantle, heat sources (derived principally from decay of radioactive material) generate convection cells and plastic flow in the asthenosphere. Upward-converging convection cells may result in upward arching of the lithosphere and zonal concentration of heat at the surface. These zones may fracture, pull apart and become sites of volcanism, which introduces new rock to the surface. Lateral flow of the asthenosphere may, in turn, drag segments of the lithosphere outward from these zones of rifting. Because heat sources apparently are not evenly distributed within the mantle, nor are they of equal magnitude, the sizes and geometries of the convection cells are variable. This results in non-uniform tectonic patterns on a global scale, in which individual segments of the lithosphere, or **plates** (Fig. 3.13), are of different sizes and are moving at different rates.

In general, zones of upward-converging convection are sites of **mafic** (magnesium- and iron-rich magmas) volcanism, which reflect the iron-rich sources of magma derived from the mantle. Such volcanism leads to the generation of new crust and commonly occurs in oceanic settings. Up-arching of the crust and accumulation of lava form a symmetric mid-ocean ridge and central rift. Lateral flow away from the central rift, termed **sea-floor spreading**, has been measured to a maximum of about 16 cm yr^{-1} along the East Pacific Rise.

Downward-converging convection cells result in drag of lithospheric plates toward one another into collision. Any one of several styles of plate collision may occur, depending upon the composition of the crustal segments that are involved and the angle of collision (head-on, orthogonal, etc.). Downward dragging of slabs of crust, or **subduction**, generates seismic disturbances (earthquakes) and possible remelting of the crust, leading to volcanism. Because the magma is generated at least partly from sial (either from continental crustal materials or from oceanic sediments), volcanism in subduction zones tends to be more **silicic** (silica rich) in comparison to zones of up-welling convection.

The global tectonic patterns, related to plate motion on Earth, provide an important model for comparing the

Figure 3.8 High-altitude oblique aerial photograph of the eastern San Francisco volcanic field, Arizona, showing prominent grabens (two parallel faults bounding a down-dropped block), cinder cones (left side) and lava flows, some of which have flowed into the grabens (photograph courtesy US Geological Survey, Flagstaff).

Figure 3.9 Aerial photograph of faulted strike ridges formed of steeply dipping strata (Lookout Ridge, Alaska), cut by high-angle cross-faults. The surrounding plains are being eroded by solifluction (US Geological Survey photograph BAR 44–47).

Figure 3.10 Oblique aerial view northwestward along the San Andreas fault in California. The fault marks the major boundary between the Pacific and the North American plates. Sheared rocks along the fault zone are preferentially eroded (photograph by Robert Wallace, US Geological Survey).

geophysical characteristics of all solid-surface planets. The morphologic expression of these tectonic patterns provides the key to the interpretation of internal processes on all planets, including the Earth.

3.3.3 Surface morphology of tectonic features

Local tectonic features—more commonly referred to as **structures**—include various faults, joints and folds, shown in Figure 3.14. Faults involve shearing of rocks and occur in three principal types: (a) **normal faults** which result from tensional stresses; (b) **reverse faults** which result from compressional forces (thrust faults are reverse faults involving very low angle fault planes); both normal and reverse faults involve primarily vertical displacements of rocks along the fault plane; and

Figure 3.11 Anticlinal fold in Silurian sandstones and shales, Washington County, Maryland (US Geological Survey photograph by I. C. Russell).

Figure 3.12 Vertical aerial photograph of Circle Ridge Dome, Fremont County, Wyoming. This eroded dome has a central core underlain by Triassic shales and sandstones surrounded by outward-dipping Jurassic sediments (US Geological Survey photograph GS IE6-32, 1948).

49

Figure 3.13 Map showing principal crustal plates on Earth (courtesy US Geological Survey).

(c) **strike-slip faults** in which displacement is principally in horizontal directions.

Joints, like faults, also involve fractures, but unlike faulting in which rocks shear along the fracture, jointing only involves separation of rocks away from the fracture. Because the faulted and jointed rocks are fractured, often they are more easily eroded and the fractures themselves may become widened by preferential erosion (Fig. 3.10).

Folds also result from crustal deformation, but rather than fracturing, as occurs in faulting and jointing, the rocks yield by bending. Folds of several geometrics may occur, some of which produce topographic expressions that directly reflect the form of the fold. Thus, in Figure 3.15 the mountains coincide with **anticlines**. More commonly, however, there is an inversion of topography: during the process of folding, rocks along the axes of anticlines are subjected to tension which may cause them to be jointed, or in effect to "open up"; conversely, rocks along the axes of **synclines** are subjected to compression, which may cause pore space and other voids to close. Thus, weathering and erosion are enhanced along the axes of anticlines and are retarded along the axes of synclines. With time, net erosion is more rapid over the anticlines, leading to topographic inversion in which valleys develop along the axes of the anticlines and the synclines remain as ridges (Fig. 3.16).

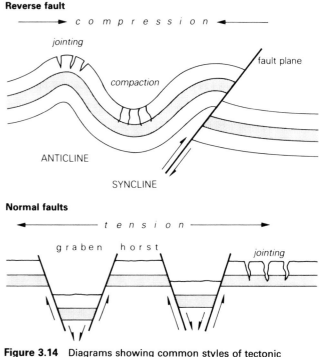

Figure 3.14 Diagrams showing common styles of tectonic deformation into faults, folds and joints and the formation of horsts and grabens.

50

Figure 3.15 Apollo 7 photograph of the Zagros Mountains, Iran viewed to the north showing active mountain-building zone in which topography mirrors the structure; most of the mountains are anticlines; dark circular areas are intrusive salt domes (Apollo 7 AS7-15-1615; after Lowman 1981).

Although individual faults, joints and folds may range in size up to a few tens of kilometers and may be related to localized tectonic or igneous events, complex systems of multiple structures can extend hundreds of kilometers as part of tectonic deformation associated with crustal plate motion (Fig. 3.17). From orbital views, only the larger individual structures and structural systems are visible, and often it is difficult to identify the specific type of structure, such as normal versus reverse faults, or normal versus inverted topography.

3.3.4 Comparative planetary tectonism

As reviewed by Head and Solomon (1981), the types of landforms exhibited on the terrestrial planets and the timing of their formation are intimately linked to thermal evolution and the subsequent planetary interior characteristics. The smaller terrestrial planets—Mercury, the Moon and Mars—have thick lithospheres which show little evidence of destruction or renewal over the last 80 percent of their history, in contrast to Earth and possibly Venus. These so-called *one-plate* planets are dominated by early-formed crusts which preserve the period of heavy bombardment and have histories that suggest ever-thickening lithospheres. Tectonism on these planets is expressed primarily by vertical movements, forming features such as grabens.

In contrast to one-plate planets, the lithosphere of Earth is highly mobile and exhibits extensive lateral movement, as discussed in previous sections. Venus, similar in size to Earth, has landforms unlike the smaller planets, but may also show evidence suggestive of large impact craters. However, until high-resolution images of Venus become available, its tectonic history remains open to considerable debate, as reviewed in Chapter 6.

3.4 Volcanic processes

Volcanic processes involve the generation of magma and magma-related materials and their eruption onto the surface. Thus, volcanic structures provide direct clues to the thermal evolution and interior characteristics of planetary objects. On Earth, magma appears to be generated in the lower crust and upper part of the mantle as a result of heating from a variety of sources, including: (a) heat generated by the differentiation of the mantle and core; (b) frictional heat generated by tectonic and body-tidal processes; and (c) heat derived from radionuclides. An examination of the location of most active volcanoes on Earth shows a more than coincidental correlation with

tectonic plate boundaries. As discussed in Section 3.3, spreading zones are typically marked by mafic volcanism, involving ferromagnesium-rich silicate magmas which typically produce basaltic rocks.

Subduction zones are also sites of volcanism and typically involve silica-rich magmas which commonly produce andesitic, dacitic and rhyolitic lavas and ash deposits (Fig. 3.18(a) and (b)). As the subducted plates are carried to greater depth, they begin to melt, and because the crustal plates are composed of lower-density materials than the mantle, the melt tends to rise. Depending upon the geometry of subduction, additional crustal materials may be incorporated in the melt during its ascent to the surface.

Volcanism may also occur in mid-plate zones, both in oceanic environments, as typified by the Hawaiian Islands, and in continental environments, represented by the Tibesti Highlands of north-central Africa. Some of these mid-plate volcanic zones are postulated to result from a rising thermal "plume" in the mantle which upwells in a fixed, centralized location. As the overlying crustal plate slides across the plume, magma periodically erupts through the plates to the surface (Fig. 3.18(c)). Thus, chains of volcanoes may be generated on the plate in an assembly line fashion in which the volcanoes are progressively older in the chain away from the source. Close correspondence of the rate of plate motion and of the ages of the volcanic rocks in the Hawaiian Emperor chain on the Pacific plate lends credence to this hypothesis (Dalrymple *et al*, 1973).

Basaltic volcanism is an extremely important process on the terrestrial planets. Basalts form the floors of the oceanic basins and have erupted on Earth throughout its known geologic history. The dark mare areas of the Moon are basaltic lava flows, constituting at least one-fifth of the lunar surface; perhaps 50 percent or more of the martian surface is covered with basaltic materials, as may be substantial parts of the plains on Mercury. Several lines of evidence suggest that the asteroid Vesta may be basaltic, and some meteorites appear to be fragments of bodies that experienced basaltic volcanism. Although speculative, some of the mountains and plains of Venus may also be the result of basaltic volcanism. The importance of basaltic volcanism in the inner Solar System prompted the formation of a project involving more than 100 scientists to consider all aspects of the topic in the planetary context. Begun in 1976, the project culminated with the publication of a key reference entitled *Basaltic volcanism on the terrestrial planets* (BVSP 1981).

Numerous classifications of volcanic eruptions have been derived. Most classifications for the *styles of volcanism* are based on the characteristics of the products erupted and on the degree of eruption explosivity, as given in simplified form in Table 3.2.

Figure 3.16 Landsat mosaic prepared by US Soil Conservation Service showing eastern USA, centered on 40°N, west of the Atlantic coast. Present ridges and valleys of Appalachians were formed by differential erosion after regional uplift, not directly by tectonism as in the Zagros Mountains shown in Figure 3.15 (after Lowman 1981).

Figure 3.17 Landsat view of Tien Shan, China, centered near 78°E, 40°N, showing folds and overthrusts; note prominent overthrust at lower left, bounded by left-lateral tear faults (Landsat image 1206-05000; after Lowman 1981).

Figure 3.18(a) Landsat image of Cerro Panizos ignimbrite shield, Bolivia/Argentina border (from Francis and Wood 1982), showing radial drainage gullies cut into ash flows and domes in the central area. Area shown is 100 km across, centered at about 23°S, 66.5°W (Landsat 1008-13531 and 2256-13411, courtesy P. Francis).

Figure 3.18(b) Mount St. Helens volcano in the Cascade Range, USA, prior to the eruptions of 1980. This composite, or strato-volcano, grew by intermittent eruptions of lavas which ranged in composition from basalt to andesite and dacite. The volcano is 10 km in diameter and reaches 2000 m in elevation; view is to the east (photograph 76-A, R. Greeley).

Figure 3.18(c) Oblique aerial view of Mauna Loa shield volcano, Hawaii (left side), composed predominantly of basaltic lava flows, and Mauna Kea (right side) a more "evolved" volcano, the summit of which is marked by cinder cones (US Navy photograph, 0066, Nov. 1954).

3.4.1 Volcanic morphology

The forms of volcanoes and volcanic terrains have been studied for many years. Cotton (1952) analyzed volcanic geomorphology as related to various processes including styles of eruption. Wentworth and Macdonald (1953), Macdonald (1967, 1972) and Green and Short (1971) also describe volcanic features. More recently, Head *et al.* (1981*b*) and Whitford-Stark (1982) have discussed factors involved in the morphology of volcanic landforms, especially as related to planetary geomorphology. Most of these references are focused on basaltic volcanism but, unfortunately, much less attention has been given to silicic volcanism.

The ultimate forms of volcanoes and related terrains are the result of many complex, often interrelated parameters. Whitford-Stark (1982) notes that these factors fall into three groups (Table 3.3): (a) planetary variables, (b) magma properties controlling rheology and (c) intrinsic properties of eruptions. Planetary variables include those factors that are characteristic for the particular body in question. For example, the height of an ejection plume in an explosive eruption would be governed by such considerations as the presence or absence of an atmosphere, the gravitational acceleration and the escape velocity. In turn, these factors influence the shape of the volcano; in an airless, low-gravity environment such as the Moon, pyroclastic deposits would be widespread, in contrast to Earth where the ejection distance would be retarded by the atmosphere and higher gravity (McGetchin and Head 1973), which could lead to the formation of cinder cones (Fig. 3.19).

Magma rheology is an extremely important parameter in volcanic morphology. Fluid lavas spread more easily, leading to the emplacement of volcanic **plains** often "fed" by **lava channels** (Fig. 3.20) and **lava tubes** (Fig. 3.21), in contrast to viscous lavas which typically form short, stubby flows accumulating as **domes** (Fig. 3.22). Many properties of the magma control the rheology of lavas, as shown in Table 3.3.

Table 3.2 Styles of volcanism.

Style	Products	Activity	Typical composition	Vent types
Flood eruption	Fluid lavas	High rate of effusion, little explosivity	Basaltic	Fissure
Plains volcanism	Fluid lavas	Moderate rates of effusion	Basaltic	Aligned central vents and fissures
Hawaiian	Fluid lavas, modest pyroclastics	Moderate rates of effusion, sporadic eruptions	Basaltic	Predominantly central vents, some fissures
Strombolian	Pyroclastics, some lavas	Low to moderate rates of effusion, sporadic	Basaltic, andesitic	Central vent
Explosive	Pyroclastics (mostly ash); modest, viscous flows	Moderate rate, but highly energetic	Andesitic, dacitic	Central vent
Rhyolitic flood	Ash flows	Very high rate of eruption	Rhyolitic	Presumably fissure

Table 3.3 Factors governing the morphology of volcanic landforms (from Whitford-Stark 1982).

Planetary variables	Magma properties controlling rheology	Properties of eruption
Gravity	Viscosity	Eruption rate
Lithostatic pressure	Temperature	Eruption volume
Atmospheric properties	Density	Eruption duration
Surface/ subsurface liquids	Composition	Vent characteristics
Planetary radius	Volatiles	Topography
Planetary composition	Amount of solids	Ejection velocity
Temperature	Yield strength shear strength	

Various characteristics of the eruption constitute the third main group of factors. Figure 3.23 illustrates how one of these factors, rate of effusion, may influence volcanic morphology. Low rates of effusion produce relatively short flows, whereas high rates of effusion produce long flows, as documented by Walker (1973). Short flows tend to accumulate close to the vent, forming lava domes and cones, in contrast to long flows which produce lava plains.

3.4.2 Volcanic craters

Volcanic craters can form through a variety of processes. Craters larger than about 2 km are termed **calderas** and can form by collapse, explosive eruptions, erosion or a combination of these processes. Typically, calderas involve multiple eruptions and as a consequence show multiple vents and complex histories (Fig. 3.24). Smaller (≤1 km) collapse features, termed **pit craters**, commonly form in basaltic lavas and, although frequently are vents, can form without associated eruptions. Still smaller craters, termed **collapse depressions** (Fig. 3.25), form on basalt flows and are not generally related to vent activity.

Maars are volcanic explosion craters which result from phreatomagmatic eruptions in which rising magma encounters water (either surface or subsurface water). Because these are "point-source" explosions, in many ways maars resemble impact craters (Fig. 3.26).

Figure 3.27 is an attempt to classify common volcanic features and to relate them to some of the more important factors involved in volcanic eruptions. Thus, for planetary purposes it is possible to approximate the styles of eruption based on the morphology of the volcanic features. Unfortunately, because so many of the factors involved in volcanic morphology are interrelated and impossible to disentangle by commonly available remote-sensing methods, the details of the volcanic processes involved in their formation usually remain unknown.

3.4.3 Intrusive structures

Not all magma reaches the surface to produce volcanoes. Some magma intrudes crustal rocks, cools and crystallizes, and may later be exposed through weathering and erosion. Frequently, these intrusive structures are more resistant to erosion than the intruded host rock and stand out as topographic features. Figure 3.28 shows Green Mountain in Wyoming, a dome of sedimentary rocks underlain by an igneous intrusion. Figure 3.29 shows several intersecting ridges composed of igneous rocks which intruded as vertical sheets termed **dikes**.

3.5 Gradation

Gradation is a complex process that begins with weathering and erosion, continues with transport of the weathered debris, and ends with deposition of the material. Thus, gradation is the "leveling off" process in which topographically high areas are worn away by erosion and low areas are filled by deposition. The driving force of gradation is gravity. Through gravity, material is moved

Figure 3.19 Cluster of small cinder cones in Hawaii; morphology of cinder cones may be highly variable as to planimetric shape and profile; some cones lack summit crater.

Figure 3.20 Lava channels up to 10 m wide on the southwest ▷ rift zone of Mauna Loa, Hawaii; the rift zone runs diagonally from upper right and is marked by a series of cinder-and-spatter cones which were the vents for flows in which the channels developed (NASA-Ames photograph by R. Greeley).

Figure 3.21 Vertical aerial photograph of the southeast rift zone of Hualalai Volcano, Hawaii, showing collapsed lava tube that originated in flows from the cinder cone on the left (US Dept of Agriculture photograph, 1965).

Figure 3.22 Oblique aerial photograph of Mono Craters, eastern California. These volcanic domes are composed predominantly of rhyolitic obsidian. On the left are two short, stubby lava flows originating from a large dome. Panum Crater, on the right, shows a central dome surrounded by a ring of pyroclastic deposits. View is to the southwest, with Mono Lake in the foreground and the Sierra Nevada mountains in the distance (US Geological Survey photograph by C. D. Miller).

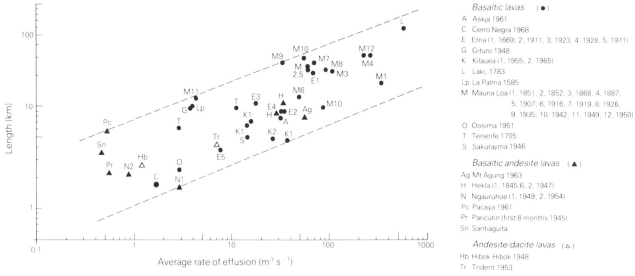

Basaltic lavas (●)

A Askja 1968
C Cerro Negra 1968
E Etna (1, 1669; 2, 1911; 3, 1923, 4, 1928, 5, 1971)
G Gituro 1948
K Kilauea (1, 1955, 2, 1965)
L Laki, 1783
Lp La Palma 1585
M Mauna Loa (1, 1851, 2, 1852, 3, 1868, 4, 1887,
5, 1907, 6, 1916, 7, 1919, 8, 1926,
9, 1935, 10, 1942, 11, 1949, 12, 1950)
O Oosima 1951
T Tenerife 1705
S Sakurajima 1946

Basaltic andesite lavas (▲)

Ag Mt Agung 1963
H Hekla (1, 1845/6, 2, 1947)
N Ngauruhoe (1, 1949, 2, 1954)
Pc Pacaya 1961
Pr Paricutin (first 8 months 1945)
Sn Santiaguita

Andesite/dacite lavas (△)

Hb Hibok Hibok 1948
Tr Trident 1953

Figure 3.23 Plot of lava flow length against average effusion rate for lava eruptions (mostly basaltic) on various volcanoes (from Walker 1973; reprinted with permission of the Royal Society).

Figure 3.24 Vertical aerial photograph of Mount Tavurur, New Britain, showing complex caldera reflecting multiple eruptions (stereogram no. 102 of the University of Illinois Committee on Aerial Photography).

by **mass wasting**—such as landslides—by running water, by frozen water (glaciers) or by wind.

3.5.1 Mass wasting

Mass wasting is the downslope movement of rock and debris under the influence of gravity (Sharpe 1968) and, thus, is a universal geologic process. Figure 3.30 classifies various forms of mass wasting. In general, mass wasting is subdivided on the basis of rate of movement, the types of material that are involved and water content. Water acts in several ways to enhance mass wasting: (a) a film of water acting as a lubricant, can destroy the cohesion between particles; (b) in many materials, particularly the clay minerals, water may enter the crystal structure causing swelling and disruption of the strength of the material; (c) water adds weight to the potential landslide masses and thus helps to "push" the mass down hill; and (d) fluid pore pressure can reduce the amount of energy necessary to initiate movement in both faults and landslides. Figure 3.31 shows various styles of mass wasting on Earth.

Features related to mass wasting have been observed on Mars, where water may have been a contributing factor, and on the Moon and Mercury, both of which lack water.

3.5.2 Processes associated with the hydrologic cycle

The **hydrologic cycle** defines the circulation of water among surface reservoirs (such as oceans), the atmosphere and groundwater systems. On Earth, water is a dominant geologic agent and the hydrologic cycle figures prominently in surface processes. Little can be said about the possible hydrologic cycles on other planets. Water does not exist on the Moon, nor has it ever existed according to models based on analyses of lunar samples, and it is highly unlikely that water has ever existed on Mercury, because of its close proximity to the Sun. Although water exists on Mars and many of the outer planet satellites, only on Earth are conditions favorable for liquid surface water. However, the presence of ancient channels and valley networks on Mars suggests that climatic conditions in the geologic past were different and that liquid water probably existed on the surface. Moreover, liquid water may exist at depth in certain regions of Mars and some of the outer planet satellites.

Venus, too, may have had liquid surface water, although present conditions far exceed the boiling point of water (Table 2.1). Interpretations based on measurements of certain isotopes in the atmosphere have led some investigators to propose oceans of water in the past (Donahue *et al.* 1982). When high-resolution radar images are obtained of the venusian surface, the search for ancient river beds and other water-related landforms will be a high priority.

With channels observed on Mars and the possibility of water having existed on Venus, it is important to assess the morphologic features associated with the hydrologic cycle, at least as seen on Earth, as a basis for comparison. Numerous texts on geomorphology and hydrology have been written about streams (Leopold *et al.* 1964 and others), but here we consider only the major features which appear to be important for planetary comparisons.

River and stream patterns provide clues to the structure and characteristics of the underlying rocks and topography. Howard (1967) provides an exhaustive classification of drainage patterns for Earth and gives possible geologic significance for each category. His classification takes into account the whole system (valleys, gullies, channels) on a regional scale and is, therefore, particularly appropriate for planetary comparisons which deal with large areas.

Table 3.4 summarizes Howard's scheme and is supplemented by diagrams shown in Figure 3.32. This classification is largely empirical and patterns may grade from one form into another. Nonetheless, the system provides a basis of comparison that is extremely useful. Streams and some associated features are shown in Figure 3.33.

One part of the hydrologic cycle involves groundwater. Erosion produced by groundwater dissolving certain

Figure 3.25(b) Oblique aerial photograph showing collapse depressions; most collapse depressions average 10 m in diameter (from Greeley & Gault 1979).

◁ **Figure 3.25(a)** Aerial photograph showing collapse depressions (arrows) on the Wapi lava field (basalt), Idaho; also shown is a pressure ridge (lower right corner); area of photograph is about 1.5 by 1.3 km (NASA-Ames photograph 878 5-1).

Figure 3.26 Oblique aerial view of crater Elegante, a 1.3 km maar crater in the Pinacate volcanic field, northern Mexico. Maar craters typically have raised rims, are shallow in relation to their diameter, and are rather circular in plan view—as such they resemble impact craters (photograph by R. Greeley, 1972).

Figure 3.27 Classification of volcanic features based on style of eruption and properties of the magma (modified from Rittmann 1962; reproduced with permission of Ferdinand Enke).

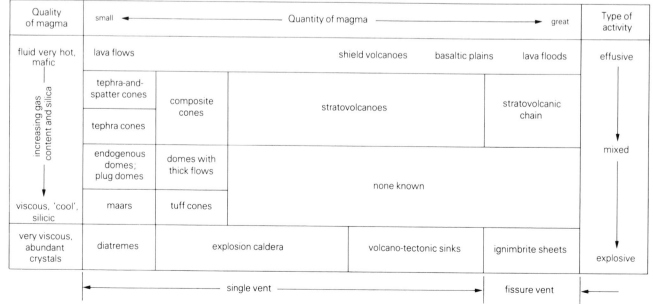

Quality of magma	small ◀─────── Quantity of magma ───────▶ great					Type of activity
fluid very hot, mafic	lava flows		shield volcanoes	basaltic plains	lava floods	effusive
↑ increasing gas content and silica	tephra-and-spatter cones	composite cones	stratovolcanoes		stratovolcanic chain	
	tephra cones					mixed
	endogenous domes; plug domes	domes with thick flows	none known			
viscous, 'cool', silicic	maars	tuff cones				
very viscous, abundant crystals	diatremes	explosion caldera		volcano-tectonic sinks	ignimbrite sheets	explosive
	◀──────── single vent ────────▶				fissure vent	

Figure 3.28 Vertical aerial photograph of Green Mountain (Crook County, Wyoming), a dome of outward-dipping sedimentary rocks underlain by a laccolith (a plutonic intrusion of igneous rocks). The dome is about 1.8 km across (US Dept of Agriculture photograph BBU-29-78).

Figure 3.29 Vertical aerial photograph of intersecting dikes, Spanish Peaks area, Colorado. Dikes of several ages are shown, indicated by cross-cutting relationships; area shown is about 3.1 by 4.5 km (US Geological Survey photograph CL-34-71).

Type of movement		Type of material		
		Bedrock		Soils
falls		rockfall		soilfall
slides		rotational	planar	planar	rotational
	few units	slump	block guide	block guide	block slump
	many units		rockslide	debris slide	failure by lateral spreading
flows		All unconsolidated			
		rock fragments	sand or silt	mixed	mostly plastic
	dry	rock fragment flow	sand run / loess flow		
			rapid earthflow	debris avalanche	slow earthflow
	wet		sand or silt flow	debris flow	mudflow
complex landslides		combinations of materials or type of movement			

Figure 3.30 Classification of landslides (modified from Varnes 1978).

rocks (typically limestones, rock salt or gypsum) leads to a terrain termed **karst topography**. Depending upon the stage of evolution, karst topography may display only a few **sinkholes** (collapse pits), or numerous sinkholes plus **solution valleys** (collapse drain network), or highly eroded karst in which only **haystacks** (Fig. 3.34), **pinnacles** and **spires** remain as erosional remnants.

Landforms with the imprint of former lakes, swamps and oceans are highly diverse. Typically, these are sites of former sedimentary deposition and, with the removal of water, leave flat, broad plains, typified by intermontane **playas**. Shoreline processes may lead to features such as **terraces** (both erosional and depositional, which may reflect former shorelines), **sea cliffs** or **beaches**. Except for some craters and canyons on Mars, which may have contained ponded water in the past (Fig. 3.35), and the possibility of former oceans on Venus, only Earth displays landforms associated with large bodies of water.

3.5.3 Aeolian processes

Any planet or satellite having a dynamic atmosphere and a solid surface has the potential for **aeolian** (wind) processes (Greeley and Iversen 1985). Most deserts, coastal areas and glacial plains and many semi-arid regions on Earth are subject to aeolian processes. Seasonal dust storms sweep across Mars where aeolian activity appears to be the dominant active process. Measurements of wind speeds on Venus and analyses of images from its surface suggest the possibility of aeolian processes. Discovery by the Voyager mission of the predominantly nitrogen atmosphere on Titan raises the possibility of aeolian activity on this satellite of Saturn.

As outlined in the classic reference by Bagnold (1941),

winds transport sediments via three modes: **suspension** (mostly silt and clay particles, i.e. smaller than about 60 μm), **saltation** (mostly sand-size particles, 60–2000 μm in diameter) and **surface creep** (particles larger than about 2000 μm in diameter). Wind threshold curves (Fig. 3.36) define the minimum wind speeds required to initiate movement of different particles for given planetary environments. The ability of wind to attain threshold is a function primarily of atmospheric density, viscosity, composition and temperature. Thus, the very low-density atmosphere on Mars (Table 2.1) requires wind speeds that are about an order of magnitude stronger than on Earth.

Aeolian processes are capable of redistributing enormous quantities of sediment over planetary surfaces, resulting in the formation of landforms large enough to be seen from orbit and deposition of windblown sediments that can be hundreds of meters thick. Because aeolian processes involve the interaction of the atmosphere and lithosphere, an understanding of aeolian activity sheds light on meteorological problems. Aeolian activity can be considered in terms of large-scale and small-scale modifications.

Large-scale modifications involve features that can be observed from distances of orbiting spacecraft. One of the most useful types of features for interpretation of surface processes is the **dune**, a depositional landform (Fig. 3.37). Both the planimetric shape and cross-sectional profile of dunes can reflect the prevailing winds in a given area (Fig. 3.38). Thus, if certain dune shapes or slopes can be determined from orbital data, local wind patterns can be determined. Repetitive viewing of the same dunes as a function of season may reveal seasonal wind patterns.

On Earth great quantities of silt and clay are transported

Figure 3.31 Various forms of mass wasting. (a, previous page, top) View of slumping in shale (Oahe quadrangle, South Dakota; US Geological Survey photograph by D. R. Crandell). (b, previous page, bottom) Talus cones resulting from rock falls and stream wash; South Stinking Water Canyon, Park County, Wyoming (US Geological Survey photograph by T. A. Jaggar, 1893). (c, above left) Rock glacier on McCarthy Creek, Copper River region, Alaska, showing source supply in talus cones and flow lobes into valley. Rock glaciers are masses of poorly sorted rocks and fine material held together by interstitial ice (US Geological Survey photograph by F. H. Moffit). (d, above right) Earthquake-induced landslide (dark flow) on the Sherman Glacier. Slide material is estimated to be 10^7 m^3 and covers an area of about 8 km^2. Tashuna district, Copper River region, Alaska (US Geological Survey photograph by Austin S. Post, 1965).

Table 3.4 Classification of drainage patterns (modified from Howard 1967).

Pattern	Significance
Dendritic	Horizontal sediments or uniformly resistant crystalline rocks. Gentle regional slope at present, or at time of drainage inception.
Parallel	Moderate to steep slopes; also in areas of parallel, elongate landforms.
Trellis	Dipping or folded sedimentary, volcanic or low-grade metasedimentary rocks; areas of parallel fractures.
Rectangular	Joints and/or faults at right angles. Streams and divides lack regional continuity.
Radial	Volcanoes, domes and residual erosion features.
Annular	Structural domes and basins, diatremes and possibly stocks.

in dust storms and eventually deposited as **loess**. Thick loess deposits are found throughout the geologic column. Even where relatively young and well-exposed on the surface, loess deposits are nearly impossible to identify as such by remote-sensing methods. Yet identification of such deposits could be very important in understanding planetary surfaces.

Large-scale aeolian erosional features include pits and hollows (called **blowouts**) that form by deflation (the removal of loose particles) and wind-sculptured hills called **yardangs** (Fig. 3.39).

Observations of active aeolian features provide direct information on the atmosphere. For example, **variable features** on Mars are surface patterns which form as a result of aeolian activity. The patterns are visible as contrasts in albedo, or surface reflectivities. Repetitive imaging shows that many of them disappear, reappear or change their size, shape or position with time. Mapping the orientations of variable features has been used to derive patterns of near-surface atmospheric circulation.

3.5.4 Glacial and periglacial processes

This section deals with planetary surface features and processes associated with cold regions and water ice. **Periglacial** refers to processes, conditions, areas, climate

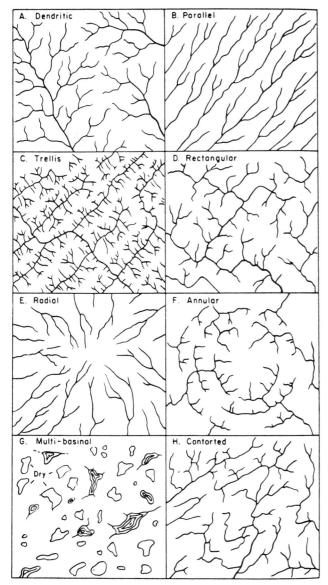

Figure 3.32 Diagram showing basic stream patterns (from Howard 1967, with permission of American Association of Petroleum Geologists).

and topographic features in cold regions or in any environment where frost action is important.

A review of the various environments in the Solar System shows that all planets and satellites except Venus experience temperatures below freezing. Subsurface ice probably exists on Mars and ice is a major constituent of many of the outer planet satellites.

Surface features on an icy or ice-rich body result largely from processes of flow and fracture. Although ice is often modeled as a Newtonian viscous fluid, experiments indicate that it can be considered a "pseudo-plastic fluid" which deforms by creep (Glen 1974). In a Newtonian fluid, the rate of strain is linearly proportional to the applied stress and the viscosity is the ratio of strain proportional to the stress raised to some power. Thus, as the

stress level is increased, the material deforms more and more rapidly. The result is that ice appears to become less viscous at higher rates of strain. However, under very rapid strain rates, such as during an impact event, ice behaves more like a brittle elastic material than a fluid.

On Earth, glaciers are classified as either **valley glaciers** (Fig. 3.40), or as **ice sheets** (also called continental glaciers, or ice caps) if they are too large to be contained by valleys. All glaciers move downslope or outward, leaving distinctive terrains (Figs. 3.41 and 3.42), and a "retreating" glacier simply means that the melting and ablation exceed the rate of forward movement by the glacier.

On Earth, precipitation of snow in the **area of accumulation** (headward part of glacier) forms a deposit that is about 20 percent ice and 80 percent air. Melting and refreezing plus compaction converts the snow to spherical ice particles called **firn**. As the firn accumulates, further compaction causes recrystallization to form the main ice mass, typically having less than 10 percent air.

Most glaciers also incorporate rocky materials within the ice mass. This material can include dust and other airborne particles and chunks of rock gouged by the ice as it moves across a surface or from the accumulation of debris derived from valley walls. Surface material often coalesces into various **moraines** which are linear deposits (Fig. 3.41). On Earth, material carried by the ice eventually reaches the front of the glacier. It may be deposited *in situ* or carried away by melt water. The finer material is often transported by the strong winds which are generated along ice margins. Coarse glacial deposits, termed **drift**, may assume a variety of geometries which provide clues to the form and position of glaciers after the ice mass has "retreated".

On Earth, glaciers have effected extensive changes in the landscape. U-shaped valleys (Fig. 3.42), grooves and striations parallel to the flow of ice, and amphitheater-shaped **cirques** in the headward parts of valleys are indicative of glacial erosion and can be seen on images obtained from orbit.

The term **periglacial** refers to a specific climatic zone in which the processes of **solifluction** (Fig. 3.43; the slow, viscous, downslope flow of water-saturated, unconsolidated materials), **gelifluction** (the flow of ice-saturated materials) and **nivation** (the erosion of rock or soil by snow and ice, by frost action and by chemical weathering) are characteristic and within which such geomorphic features as permanently frozen ground (**permafrost**), **patterned ground** (Fig. 3.44), **pingos** (Fig. 3.45) and **thermokarst** topography are developed. The occurrence of a periglacial region is not genetically related to the proximity of glaciers or continental ice sheets, contrary to what is implied by its etymology. However, the presence of water is essential for most periglacial

Figure 3.33(a) Streams and stream-related features. Headward erosion of streams, San Bernardino Mountains, California (US Geological Survey photograph by J. R. Balsley, 1949).

Figure 3.33(b) Oblique aerial photograph looking south toward the junction of the Yukon and Koyukuk Rivers, Alaska, showing meander loops, former channels and flood plain deposits (US Geological Survey photograph by US Army Air Corps, 1949).

Figure 3.33(c) Alluvial fans spreading into intermountain basins, Mojave Desert, California (US Geological Survey photograph by J. R. Balsley, 1949).

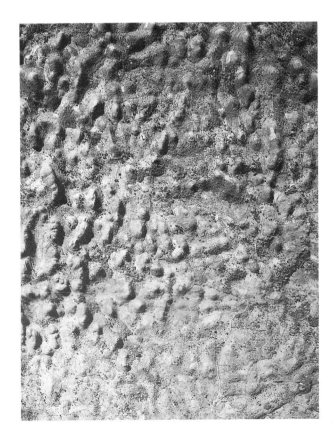

Figure 3.34 Vertical aerial photograph of karst terrain in Puerto Rico showing haystack remains resulting from solution of limestone (stereogram no. 170 of the University of Illinois Committee on Aerial Photography).

Figure 3.35 Mosaic of Viking orbiter pictures of the Lunae Planum region of Mars showing flat, dark floor of a 30 km diameter crater that appears to have been flooded with deposits associated with channels, as described by Baker and Kochel (1979) and Lucchitta and Ferguson (1983) (part of US Geological Survey subquadrangle MC-10 EC).

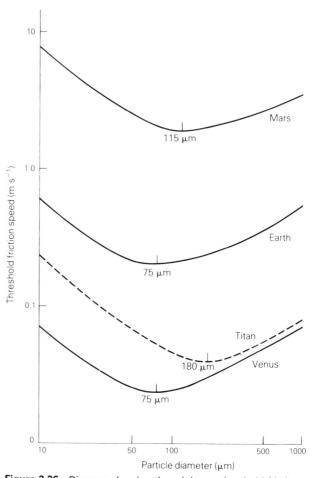

Figure 3.36 Diagram showing the minimum threshold friction speed (a function of wind speed) required to move particles of different sizes on Mars, Earth, Titan and Venus; note that as the atmospheric density decreases from Venus to Mars, minimum winds needed to set particles into motion increases.

processes to occur. This broader definition is useful in that it allows us to consider the possible operation of periglacial-type processes on the surfaces of other objects in the Solar System.

Extensive periglacially modified plains have been inferred for Mars, particularly in the northern latitudes (Carr and Schaber 1977). This deduction is based on observations and interpretations of mass wasting, some types of polygonally patterned ground and the radially striated, apparently fluidized ejecta blankets surrounding many craters.

Voyager provided a wealth of information about the outer planet satellites. With the exception of Io and possibly Amalthea, all are considered to have complex crusts consisting of mixtures of water ice and silicates. Thus, the possibility exists for mass wasting and processes of surface modification similar to terrestrial glacial and periglacial terrain. These processes may occur in association with other ices such as methane, ammonia and their clathrates. For example, the surface of Titan may consist, at least in part, of frozen methane. On Mars, deposits of carbon dioxide frost occur in the annual polar caps.

3.6 Summary

Planetary surfaces are shaped and modified by four principal processes: (a) tectonism, (b) igneous activity, (c) gradation and (d) impact cratering. Each of these processes produces distinctive landforms on Earth where most of these processes have been studied in detail. One of the goals of planetary geology is to determine how these landforms might be different in extraterrestrial environments.

Views of the Earth obtained from orbit show that while some processes can be identified by remote-sensing techniques, others cannot, thus introducing uncertainties into interpretations of planetary histories. Furthermore, with increased knowledge of the outer planet satellites, planetary geologists must assess the validity of applying Earth analogs to bodies composed mostly of ice and having markedly different environments.

68

Figure 3.37 Diagrams illustrating principal types of sand dunes and the wind direction(s) (arrows) responsible for their formation: (a) barchan, (b) parabolic, (c) transverse, (d) dome, (e) star, (f) linear (from McKee 1979).

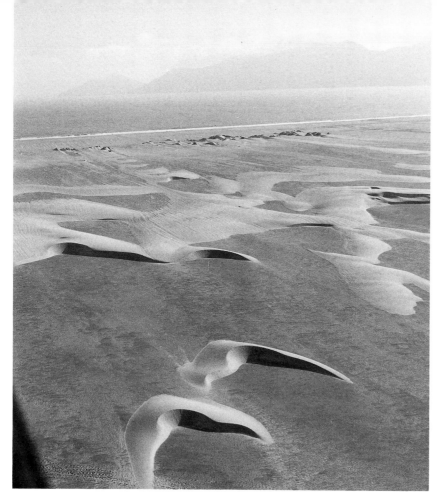

Figure 3.38 Sand dunes on Earth. (opposite) Oblique aerial view of barchan dunes in Peru; prevailing wind is from the upper left; prominent *slip faces* are on the downwind side of the dunes, toward the camera (US Geological Survey photograph by E. C. Morris). (below) Complex transverse dunes in the Algodones dune field, southern California, prevailing wind is from the left.

Figure 3.39 Wind-sculpted hills, termed yardangs, several hundred meters long, Peru (US Geological Survey photograph by J. McCauley).

Figure 3.40 Oblique aerial photograph of the Cook Inlet Region, Alaska, showing valley glaciers; dark stripes represent rocks and debris carried along with the ice (US Geological Survey photograph by A. Post, 1970).

Figure. 3.41 Vertical aerial photograph showing numerous, nearly equally spaced ridges (moraines) in Quebec; each ridge is 5–7 m high and is considered to result from debris pushed up near the front of advancing ice (from Royal Canadian Air Force photograph, prepared by the University of Illinois Committee on Aerial Photography, Stereogram No. 535).

Figure 3.42 View of Deadman Canyon, Tulare County, California, showing typical U-shaped glaciated valley (US Geological Survey photograph by F. E. Mathes, 1925).

Figure 3.43 Solifluction lobes in the Seward
Peninsula region, Alaska (US Geological
Survey photograph).

Figure 3.44 Oblique aerial view of raised-edge ice wedge polygons on the Alaskan sea coast near Barrow; polygons are 7–15 m
across (US Geological Survey photograph by R. I. Lewellen).

Figure 3.45 View of collapsed *pingo* (soil-covered ice mound) on Mackenzie Delta near Tuktoyaktuk in northwestern Canada. Long dimension of pingo is about 110 m (photograph courtesy of T. L. Péwé, Arizona State University).

4 The Moon

4.1 Introduction

The Moon has long been an object of fascination to everyone from casual observers to scientists. Aside from Earth, it is the planet about which we know most. The Moon has been photographed, probed and analyzed from orbit to determine its surface composition. Nearly a half ton of it has been returned to Earth for analyses by every conceivable method. More missions have been flown to the Moon than to any other object (Table 4.1). As discussed in Chapters 1 and 2, studies of planetary geomorphology began with the Moon. Thus, the Moon served as a 'training ground' for learning how to study the geology of extraterrestrial worlds. It is fortunate that the Moon played this role because, in many respects, it is a relatively simple object for analysis. Lacking an atmosphere, none of the complicating 'imprints' of wind or running water are superimposed on its surface. Consequently, the Moon was a far easier subject for developing methods to analyze planetary objects than if the first planet studied had been a more complicated body such as Mars.

Furthermore, the Apollo and Luna landings provided the opportunity to field-check some of the more critical relationships and interpretations that were based on remote sensing. The information from these landings enabled not only a better understanding of the Moon, but also an assessment of the techniques for deriving geologic data by remote sensing. Moore *et al.* (1980) provide a review of remote sensing applied to the Moon. Heiken *et al.* (1991) has given an excellent review of lunar geoscience.

The mid-1960s to mid-1970s saw the greatest efforts in lunar exploration through the acquisition of high-resolution images and other remote-sensing data, along with samples and results from the lander missions as reviewed by El-Baz (1979) and others. From this enormous data set, it has been possible to synthesize the general geologic history of the Moon, especially as related to the evolution of the surface. However, because of the orbital constraints placed on the Apollo missions, the equatorial regions are far better documented than the higher latitudes, which must await future missions before the knowledge for the Moon is globally uniform.

4.2 General physiography

Even with the naked eye, it is readily apparent that the surface of the Moon consists of different terrains (Fig. 4.1). The original two-fold division into the dark (albedo 5–8 percent), flat **maria** and the light-toned (albedo ~ 9–12 percent), rugged highlands or **terrae,** first described by lunar observers in the 16th and 17th centuries, is still valid, but these terrains can be further divided. As shown in Figure 4.2, five major terrains can be recognized on the Moon: (a) heavily cratered terrains, (b) moderately cratered terrains, (c) terrains associated with older impact basins, (d) terrains associated with the younger impact basins and (e) mare regions. In addition to these major units and the surface features that characterize them, impact craters are ubiquitous and important features on the Moon. Ranging in size from the microcraters observed on lunar samples to impact basins exceeding 1000 km in diameter, impact craters and related features dominate all regions of the Moon.

4.2.1 Cratered terrains

The highlands consist mostly of cratered terrains, which can be subdivided into two units based on the frequency of impact craters. In general, cratered terrain is characterized by rugged relief and high albedo and consists primarily of impact ejecta deposits and remnants of crater rims (Fig. 1.2). Models of the formation and evolution of the Moon (Taylor 1982) suggest that cratered terrain consists of highly brecciated rock masses derived from a chemically differentiated, anorthositic (rich in calcium feldspar) lunar crust. Most lunar investigators consider this differentiation to have occurred between 4.3 and 4.2 eons. During this phase of lunar history, heavy impact bombardment continuously churned the evolving lunar crust, leading to the rugged terrain seen today.

4.2.2 Basins and basin-related terrains

Basins on the Moon are defined as impact structures larger than 220 km in diameter and typically displaying concentric rings (Fig. 2.5). Studies of the oldest surfaces

Table 4.1 Lunar missions.

Spacecraft	Encounter date	Mission	Encounter characteristics
Luna 2*	12 Sept. 1959	Hard lander	Impacted lunar surface.
Luna 3*	10 Oct. 1959	Flyby	First (indistinct) photos of farside of Moon.
Ranger 7	31 July 1964	Hard lander	4300 high-resolution images with about 2000 times better definition than Earth-based photography; impacted in Mare Cognitum.
Ranger 8	20 Feb. 1965	Hard lander	7100 images obtained; impacted in Mare Tranquillitatis.
Ranger 9	24 Mar. 1965	Hard lander	5800 images obtained with resolution up to 1 m. Impacted eastern floor of crater Alphonsus.
Zond 3*	20 July 1965	Flyby	System test: took 28 pictures during flyby of Moon, then flew as far as orbital path of Mars.
Luna 9*	3 Feb. 1966	Lander	Soft landing in western Oceanus Procellarum; returned pictures.
Luna 10*	3 Apr. 1966	Orbiter	First object to orbit Moon; measured lunar magnetism and radiation.
Surveyor 1	2 June 1966	Lander	Soft landing in Oceanus Procellarum; transmitted 11,240 TV images.
Lunar Orbiter 1	14 Aug. 1966	Orbiter	Obtained 216 images, including 11 of the lunar farside. Medium-resolution pictures good, high-resolution smeared.
Luna 12*	25 Oct. 1966	Orbiter	Transmitted ~15 m resolution pictures.
Lunar Orbiter 2	10 Nov. 1966	Orbiter	Obtained 409 images.
Luna 13*	24 Dec. 1966	Lander	Soft landing in western Oceanus Procellarum; transmitted pictures.
Lunar Orbiter 3	8 Feb. 1967	Orbiter	Obtained 290 images.
Surveyor 3	20 Apr. 1967	Lander	Soft landing in Oceanus Procellarum; returned 6300 pictures; analysis of surface.
Lunar Orbiter 4	8 May 1967	Orbiter	Obtained 546 images, covering all of nearside and 95% of farside of Moon.
Lunar Orbiter 5	5 Aug. 1967	Orbiter	418 images; completed high altitude farside photographic coverage.
Surveyor 5	11 Sept. 1967	Lander	Soft landing in Mare Tranquillitatis; alpha backscatter device indicated basaltic character of that mare surface; transmitted 19,120 TV pictures.

Table 4.1 *Continued.*

Spacecraft	Encounter date	Mission	Encounter characteristics
Surveyor 6	10 Nov. 1967	Lander	Soft landing in Sinus Medii; alpha backscatter data indicated basaltic composition of surface; transmitted 29,950 TV pictures.
Surveyor 7	10 Jan. 1968	Lander	Last Surveyor; soft landing on Tycho ejecta blanket; alpha backscatter data; 21,040 TV pictures transmitted.
Luna 14*	10 Apr. 1968	Orbiter	Analysis of gravitational field.
Zond 5*	18 Sept. 1968	Flyby	Circumlunar; returned to Earth and was recovered in Indian Ocean; precursor to manned mission(?).
Zond 6*	13 Nov. 1968	Flyby	Circumlunar; returned to Earth and was recovered; precursor to manned mission(?).
Apollo 8	24 Dec. 1968	Orbiter	First manned Apollo flight to the Moon; system test of 10 orbits around the Moon; 864 photographs.
Apollo 10	21 May 1969	Orbiter	Apollo Lunar Module system test: Lunar Module descended to within 15 km of surface; 1319 photographs, color TV transmission.
Apollo 11	20 July 1969	Lander	First manned lunar landing: soft landing in Mare Tranquillitatis; 1359 photographs, 22 kg of samples returned.
Zond 7*	11 Aug. 1969	Flyby	Circumlunar; returned to Earth and was recovered; precursor to manned mission(?); 33 photographs.
Luna 16*	20 Sept. 1970	Lander	Soft landing in Mare Fecunditatis; 100 g of sample returned to Earth.
Zond 8*	24 Oct. 1970	Flyby	Circumlunar; returned to Earth and was recovered; precursor to manned mission(?); 108 photographs.
Luna 17*	17 Nov. 1970	Lander	Soft landing in western Mare Imbrium; Lunokhod I roving surface vehicle traversed 20 km.
Apollo 12	19 Nov. 1969	Lander	Soft landing in Oceanus Procellarum; 1.4 km traverse; investigated the Surveyor 3 spacecraft, deployed scientific experiments; 1577 photographs; 34 kg of samples returned.

Table 4.1 *Continued.*

Spacecraft	Encounter date	Mission	Encounter characteristics
Apollo 14	5 Feb. 1971	Lander	Soft landing near Fra Mauro crater; 3.5 km traverse; deployed scientific experiments; 1324 photographs; 43 kg of samples returned.
Apollo 15	30 July 1971	Lander	Soft landing at Hadley Rille-Apennine Mountains; 28 km (lunar rover) traverse; deployed scientific experiments; 5099 photographs plus 3375 mapping photographs (from orbit); 77 kg of samples returned.
Luna 19[*]	1 Oct. 1971	Orbiter	Orbiter only. Returned pictures.
Luna 20[*]	18 Feb. 1972	Lander	Soft landing in Apollonius Highlands; 30 g of samples returned to Earth.
Apollo 16	21 Apr. 1972	Lander	Soft landing near Descartes crater; 27 km (lunar rover) traverse; deployed scientific experiments; 4250 photographs plus 3480 mapping photographs (from orbit); 95 kg of samples returned.
Apollo 17	11 Dec. 1972	Lander	Soft landing in Taurus-Littrow Valley; 30 km (lunar rover) traverse; deployed scientific experiments; 5807 photographs plus 4710 mapping photographs (from orbit); 111 kg of samples returned.
Luna 21[*]	16 Jan. 1973	Lander	Soft landing in eastern Mare Serenitatis; Lunokhod II roving surface vehicle traversed 30 km.
Mariner 10	Nov. 1973	Flyby	Images of the north polar area.
Luna 23[*]	2 Nov. 1974	Lander	Soft landing in Mare Crisium; sample return malfunctioned.
Luna 24[*]	19 Aug. 1976	Lander	Soft landing in Mare Crisium; 160 cm core sample returned.
Galileo	7 Dec. 1990	Flyby	Multispectral images of the western limb and farside.
Galileo	7 Dec. 1992	Flyby	Multispectral images of the north pole and eastern limb.

[*]Soviet missions.

on the Moon (CLHC 1979) suggest that basins formed over an extended period prior to 3.8 eons and that the earlier basins have been mostly obliterated (Carr 1983).

Basins and basin-related terrains are subdivided into two units on the basis of age: old areas that pre-date the formation of the Nectaris basin and younger, post-Nectaris areas and basin-related terrains. Basin geology dominates the lunar surface in several ways, as reviewed by Howard *et al.* (1974): (a) all of the named mountain ranges on the Moon are segments of basin rims; (b) basins appear to have provided the structural focus for the emplacement of the mare lavas; and (c) basin ejecta deposits drape across much of the pre-basin surface on the Moon. Because of the important role of basins in lunar geology, many of the Apollo and Luna missions were specifically designed and targeted to address questions related to basin formation and evolution.

The Imbrium basin, first described by G. K. Gilbert in 1893 as an impact structure, is particularly important because it dominates much of the northern half of the lunar nearside. Gilbert recognized the distinctive radial grooves and furrows, which he termed Imbrium sculpture (Fig. 2.2), and attributed the markings to impact processes. The Imbrium basin consists of at least three rings and, although various authors have used different criteria to define these rings, Dence (1976) and Spudis (1982) suggest that the main ring is composed of the Apennine Mountains (Figs. 4.3 and 4.4), the Alpes, part of the Sinus Iridum rim, and smaller, isolated mountains which collectively define a ring about 1140 km across. Samples from the Apennines collected during Apollo 15 show that this ring consists predominantly of breccias composed of fractionated igneous rocks—mostly norites and noritic melt rocks. The intermediate ring is defined by isolated massifs such as Mons La Hire and Montes Archimedes and has a diameter of about 850 km. The inner ring is about 570 km across and is identified only by a series of ridges on the mare surface which are interpreted to reflect an underlying mountainous ring now completely buried by lavas.

Ejecta deposits originating from the Imbrium basin are spread over much of the lunar surface (Fig. 4.5). Mapped as the Fra Mauro Formation, the ejecta may be 1 km thick as far as 600 km from the basin, as determined at Julius Caesar, an impact crater that has been partly filled by ejecta from Imbrium. Because it is so widespread and can be recognized on photographs, the Fra Mauro Formation serves as a critical index formation, or datum plane, in lunar photogeologic mapping. Samples returned from the Apollo 14 mission show that the Fra Mauro Formation consists of highly brecciated rocks. Radiometric dates obtained from these samples place the age of formation for the Imbrium basin at 3.85 eons.

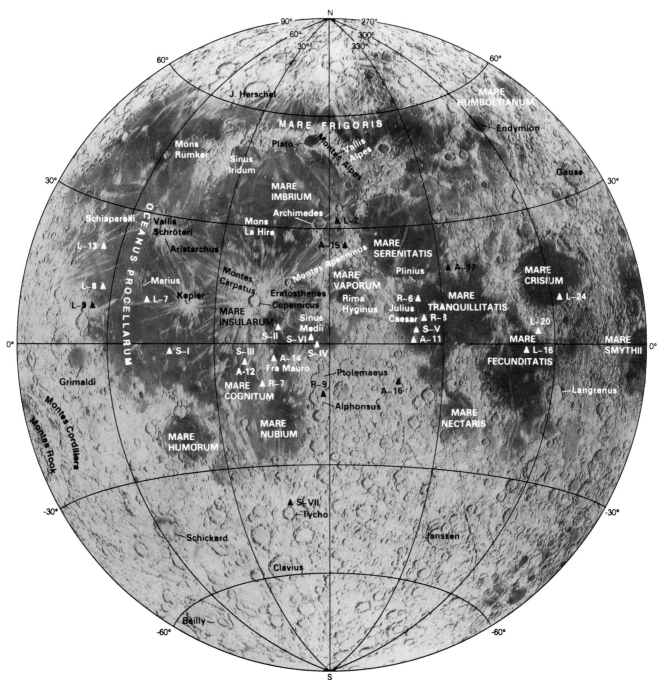

Figure 4.1 Shaded airbrush chart of the lunar nearside (a, above) and farside (b, right) showing prominent physiography and named features (base maps courtesy US Geological Survey).

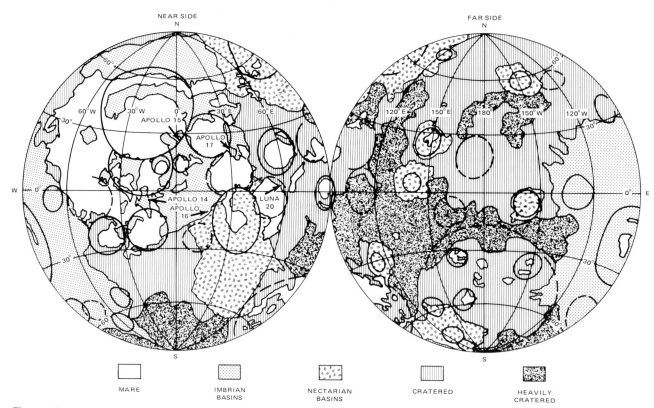

Figure 4.2 Diagram showing the distribution of the five main terrain types on the Moon (from Howard *et al.* 1974; *Rev. Geophys. Space Phys.,* copyright the American Geophysical Union).

Figure 4.3 Earth-based telescopic view of the northwest quadrant of the Moon showing the Imbrium basin (Lick Observatory photograph).

Figure 4.4 Lunar orbiter photograph showing the southeast portion of the Imbrium basin. The Apennine Bench is a structural high near the crater Archimedes (A). The light plains (AB) constitute the Apennine Bench Formation, a volcanic basin-fill unit. The Montes Apennines (MA) comprise the main basin rim in this region. Apollo 15 landing sites shown by arrow. Illumination from the right (from Spudis 1982; Lunar Orbiter IV-M109).

Detailed geologic mapping of the Imbrium basin and analysis of the orbital geochemical data have been combined with results obtained from lunar samples to synthesize its general geology. From this analysis, Spudis (1982) has derived a model of basin formation (Fig. 4.6) in which the transient crater cavity was a feature considerably smaller than the main ring, enlarged by slumping. Fracturing associated with the impact as well as various post-impact crustal adjustments was extensive and appears to have aided the eruption of early-stage lavas following the excavation of the basin. For example, Spudis (1978) has suggested that the Apennine Bench Formation (Fig. 4.4) consists of lavas emplaced as a result of these adjustments prior to the main eruptions of mare lavas that flooded the basin interior.

The Orientale basin (Fig. 2.5) is the youngest large structure on the Moon. Superposition of its ejecta, termed the Hevelius Formation (Fig. 4.7), over the Fra Mauro Formation shows that it post-dates the Imbrium basin. The Orientale basin consists of at least four rings, the

Montes Cordillera which forms a ring about 900 km across, the outer Montes Rook, the inner Montes Rook and an inner bench (unnamed). The basin is only partly flooded by mare lavas (Fig. 4.8) and hence many of the basin's interior structures are visible and can be analyzed.

Geologic mapping of Orientale (McCauley 1977) divides the pre-mare basin units into several formations (Fig. 4.9). The Maunder Formation lies between the inner and outer Rook Mountains and is a fissured, high-albedo unit interpreted to be impact melt draped over fractured blocks of basin floor (Fig. 4.10). The Montes Rook Formation lies primarily between the outer Rook Mountains and the Cordillera Mountains, although some parts of the formation extend beyond the Cordilleras. It consists of widely spaced 10 km knobs set in a background of undulating terrain and is also considered to be impact melt, but may include non-melted ejecta emplaced late in the cratering process (Scott *et al.* 1978). Most researchers favor the outer Rook Mountains defining the Orientale impact transient cavity. This conclusion is based partly

Figure 4.5 Lunar orbiter image of the Fra Mauro Formation (F) gradational with the Apennines material (AP). Framelet width is 12 km (from Spudis 1982; Lunar Orbiter IV-H109).

Figure 4.6 Sequential diagram illustrating the formation of the Imbrium basin; Dtc, diameter of transient cavity; P, peak-ring; I, intermediate-ring, A, Apennine ring; CA, Caucasus platform massif (from Spudis 1982).

on the observation that Orientale ejecta-related features occur exterior to this scarp. The Cordillera Mountains are taken to represent a "mega-terrace" by this theory. The mega-terrace is thought to have formed by the inward slump of material as the transient cavity grew, then collapsed, resulting in an enormous ring fault. Alternatively, Hodges and Wilhelms (1978) proposed that the Cordilleras represent the transient cavity. Morphologic support for this model is provided by the knobby texture of the Montes Rook Formation, which mostly occurs within the Cordilleran scarp and strongly resembles the knobby texture of the floors of medium to large ($\sim$ 50–100 km diameter) craters. The presence of ejecta-related features on this surface is explained by the assumption that high-angle ejecta falls back onto this location after the floor units are in place. Mountain rings and scarps interior to the transient crater rim, regardless of location, are thought to result from isostatic rebounding and rebound induced by impact energy.

Most lunar geologists have noted that there are signifi-

cant differences among the morphologic and structural features of lunar basins (CMRB 1980). A detailed study of five basins on the Moon which span a wide range of ages suggests to Spudis (1982) that many of the observed morphologic differences can be attributed to impacts which occurred in an ever-thickening lithosphere with time. Drawing upon models of thermal evolution and estimates of lithospheric thickness based on geophysical and geochemical considerations, Spudis shows how the differences in the morphology among various lunar basins can be accounted for, as depicted in Figure 4.11.

4.2.3 Highland plains

A widespread unit found in many highland areas consists of a high-albedo, relatively smooth, flat plains unit (Fig. 4.12). First mapped by Dick Eggleton of the US Geological Survey as the Cayley Formation, this unit was considered to be ejecta from one of the late-stage impact basins.

Figure 4.7 Orientale ejecta comprising the Hevelius Formation is shown here overriding craters and leaving coarsely festooned deposits. The floor of the crater in the lower left is buried by "deceleration dunes" where the ejecta appears to have encountered a topographic obstacle (the far crater wall) and "stalled". This area is 1100 km southeast of the center of the Orientale basin and ejecta movement was from the upper left to lower right. Inghirami, the crater in the lower left, is 91 km in diameter (Lunar Orbiter IV-H172).

The Cayley plains were later reinterpreted by some investigators as representing eruptions of volcanic ash or flows, possibly of silicic compositions. Because the unit is widespread over much of the Moon and because of its possible volcanic origin, it was given high priority as an Apollo landing site.

Samples of the Cayley Formation returned by the Apollo 16 astronauts were found to consist predominantly of impact-generated breccias, confirming its impact origin at that site. The unit is now widely regarded as ejecta deposits emplaced either as material thrown out during major basin-forming events or as local material ejected by the secondary craters of basin-forming impacts. The smooth texture of the Cayley and other Imbrium-age light plains units is attributed to the liquid-like behavior of the unconsolidated ejecta deposits when subjected to seismic shaking or acoustic fluidization.

More recent studies, however, have generated new interest in the possibility of volcanism during the early history of the Moon. Ryder and Taylor (1976) and Ryder and Spudis (1980) have synthesized data based on lunar samples and concluded mare volcanism was active prior to the formation of the Imbrium basin, possibly earlier than 4.2 eons. Studies by Schultz and Spudis (1979) of the distribution of dark halo craters > 1 km in the highlands (primarily around Mare Humorum and in the eastern hemisphere) also suggest early-stage volcanism. They propose that these impact craters excavated early-stage mare-type larvas that are presently buried by ejecta deposits from large craters and basins.

4.2.4 Maria

The maria have generally received much more attention than the highlands, despite the fact that mare areas constitute much less of the lunar surface than the highlands. It is important to note that **mare** and **basin** are not the same (see Stuart-Alexander and Howard (1970) for a review); while it is true that most maria occupy the low-lying terrain of basins, the dark maria are lava flows that were erupted generally long after the formation of most of the impact basins. Mare materials also flood some irregular depressions, such as Tranquillitatis and Procellarum, and occur in isolated patches within the highlands and on some crater floors.

Head (1976) estimates that mare lavas cover about 6.3×10^6 km², or about 17 percent of the lunar surface. Although thicknesses of mare deposits are difficult to determine, estimates by DeHon (1979) and Hörz (1978) based on the degree of flooding of impact craters indicate that mare lavas may be 4 km thick in some areas. Combin-

Figure 4.8 The west side of the Moon seen under nearly full illumination. Note the sparsity of mare flooding in the Orientale basin (arrow). The Rock and Cordillera scarps appear as bright rings under the high sun (USSR Zond 8 photo 12-306, 2525 H).

Figure 4.9 Geologic sketch map of the Orientale basin area. The units of the Orientale Group are as follows: (a) Maunder Formation, (b) Montes Rook Formation, knobby facies, (c) Montes Rook Formation, massif facies, (d) Hevelius Formation, inner facies, (e) Hevelius Formation, outer facies, (f) Hevelius Formation, transverse facies (deceleration dunes), and (g) Hevelius Formation, secondary crater facies (from McCauley 1977).

ing these estimates with their total surface area shows that mare units comprise less than 1 percent of the total volume of the lunar crust.

Analysis of mare samples returned from the Apollo and Luna missions shows that lunar lavas are very similar to terrestrial basalts but lack hydrous minerals and have slightly higher abundances of iron, magnesium and titanium. The lack of hydrous minerals is an indication that the Moon has never experienced liquid water on the surface, nor has had an appreciable atmosphere. Analysis of the samples also indicates that most of the lunar lavas were derived from deep within the mantle, probably at depths of 150 to as much as 450 km.

Crystallization dates for the cooling lunar lavas show that most of the mare lavas were erupted during the interval of 3.9 to 3.1 eons. Thus, for a period of more than 800 million years, the Moon experienced extensive volcanism. While this interval of time is less than one-eighth of the total lunar history, it nonetheless represents a span equivalent to the Phanerozoic eon on Earth. Furthermore, analysis by Schultz and Spudis (1983) indicates that both early-stage (>4.0 eons) and very late (~ 1 eon) volcanism probably occurred, extending the range of lunar volcanic activity over a 3×10^9 year period.

Laboratory studies of simulated lunar lavas provide insight into their physical properties. For example, esti-

mates of the viscosities yield values of about 10 poise, or the equivalent to motor oil at room temperature, while studies of their thermal conductivities show that heat loss would have been minimal. Thus, the lunar lavas were extremely fluid and were able to flow long distances on relatively gentle gradients.

Photogeologic mapping has been combined with terrestrial analog studies and compositional distributions based on remote sensing to derive general models for the emplacement of the mare lavas. At least three eruptive phases can be recognized. Early-stage eruptions involved lavas high in titanium recognized by their comparatively blue spectra, which erupted at high rates to form thick, massive cooling units which flooded low-lying areas such as basin centers. These units formed vast lava lakes, seen today as flat plains which generally lack surface features other than mare ridges (described below). The early-stage lavas probably required hundreds or possibly thousands of years to cool and solidify (BVSP 1981).

The first phase was followed by the eruptions of lavas having a lower titanium content, as indicated by their relatively redder spectral characteristics. These lavas appear to have been emplaced at lower rates of effusion than the first phase, suggested by formation of various lava channels and lava tubes seen today in the form of lunar sinuous rilles (Fig. 3.7). The final stage involved

85

Figure 4.10 Lunar orbiter photograph of the inner part of the Orientale basin; smooth plains (S) and fissured deposits (F) comprise the Maunder Formation considered to be impact melt, which grades into the knobby Montes Rook Formation (R). Mare lavas are visible at the top of the photograph; framelet width is 12 km (Lunar Orbiter IV-H195).

Figure 4.11 Series of diagrams illustrating the formation of five major basins on the Moon. In this model Spudis considers the differences in morphology to result from an ever-thickening lithosphere with time; thus, the formation of Crisium penetrated a relatively thin lithosphere with rapid isostatic adjustment. The sequence progresses through to the formation of the Orientale basin which impacted a relatively thick lithosphere and the attendant adjustment was followed by little extrusion of lavas in the interior (courtesy P. Spudis, 1992).

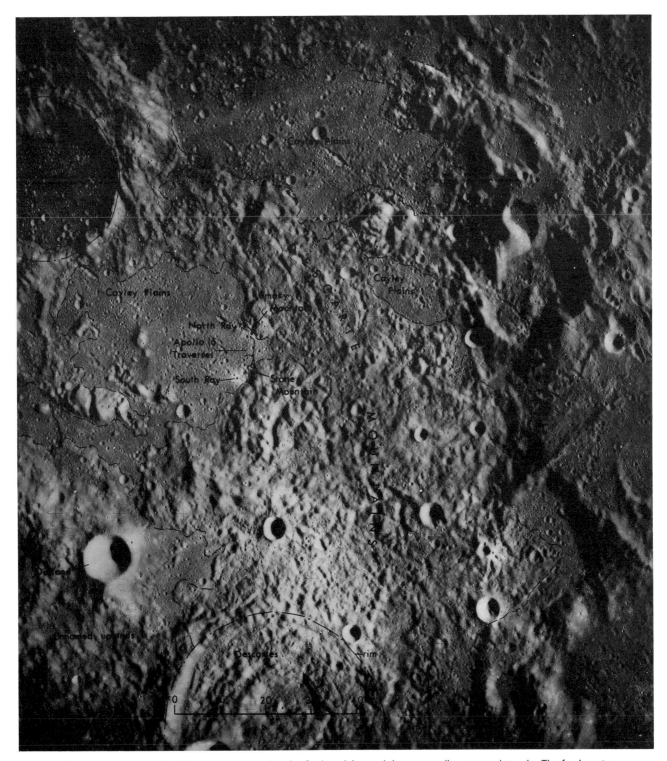

Figure 4.12 View of the Apollo 16 landing site showing the Cayley plains and the surrounding cratered terrain. The fresh crater (Dolland) in the lower left is about 10 km in diameter (AS16-M-439; courtesy C. Hodges).

87

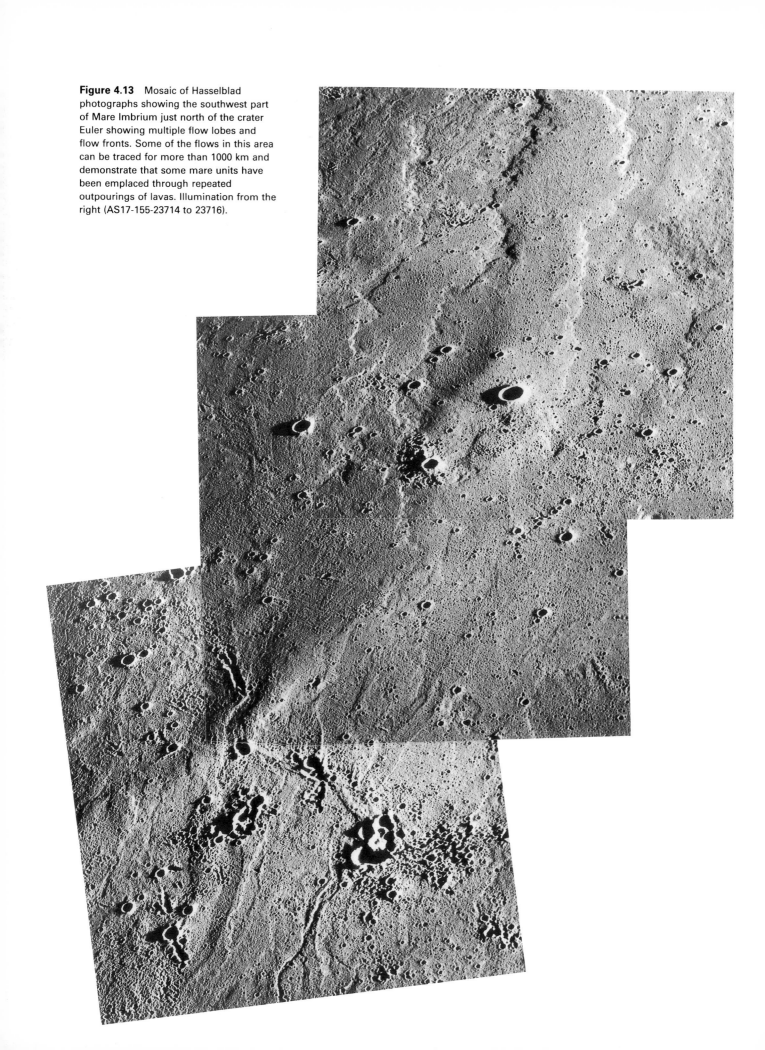

Figure 4.13 Mosaic of Hasselblad photographs showing the southwest part of Mare Imbrium just north of the crater Euler showing multiple flow lobes and flow fronts. Some of the flows in this area can be traced for more than 1000 km and demonstrate that some mare units have been emplaced through repeated outpourings of lavas. Illumination from the right (AS17-155-23714 to 23716).

lavas high in titanium, like the first phase, but erupted in lower volumes. Detailed mapping has shown that local sequences are much more complicated than this generalization.

The sources or vent areas for most lunar lavas are difficult to identify. In many cases, particularly the early-stage high-titanium lavas, the vents are presumed to have been fissures which the flood eruptions and later-stage lavas have buried. Mapping by Schaber (1973) of the well-known Imbrium lavas (Fig. 4.13) shows that they originated near the zone defined by the outer basin ring and then flowed toward the center of the basin. Some of these flows exceed 1200 km in length and have flow fronts that are 10–65 m high. Thus, the flows formed rather thick units, despite their apparent fluidity.

Visible flow fronts and margins, however, are rare on the Moon. Either most lavas were too fluid to preserve flow fronts large enough to be visible, or they have been eroded by numerous small impacts. The margins of mare lavas in several areas, however, show distinctive benches which are interpreted to represent high stands of lava (Fig. 4.14).

Several areas on the Moon appear to be volcanic complexes which served as source areas for some of the mare lavas (Guest 1971, Guest and Murray 1976). The Aristarchus Plateau (Fig. 4.15), the Marius Hills (Fig. 4.16) and the Rümker Hills (Fig. 4.17) all appear to be areas that have experienced repeated volcanic activity.

4.2.5 Sinuous rilles

Lunar *sinuous rilles* occur in many areas on the Moon, although most are found around the outer margins of the mare-filled basins (Fig. 3.7). The origin of these features has been controversial, but most investigators now agree that they are former lava channels and/or collapsed lava tubes. As such, lunar sinuous rilles can be used to map the general source areas for some of the mare lavas and to give an indication of the flow directions that were involved in the emplacement of the lavas. Although lunar sinuous rilles are typically an order of magnitude longer and wider than their presumed counterparts on Earth, models based on estimated differences in the rate of eruption by Wilson and Head (1981) and considerations of differences in gravity show that some of the differences in size (compared to Earth) can be accounted for by the lunar environment.

The Apollo 15 astronauts visited the Hadley Rille, a prominent lunar sinuous rille that was one of the sources for the emplacement of lavas into the eastern part of the Imbrium basin (Greeley 1971, Howard *et al.* 1972; Fig. 4.18). This rille stretches more than 130 km and originates in a cleft-shaped depression about 10 km long by 5 km

Figure 4.14 Topographic bench (arrow) along mare-highland contact interpreted as a high lava mark or terrace. Lavas may have been locally drained in this region by the sinuous rille, lowering the surface. Oblique view southward; near field width approximately 30 km, illumination from the left (AS16-19140; from Greeley & Spudis 1978a).

wide at the base of the Apennine Mountains. It was a vent fed by magma erupted to the surface apparently in association with basin-related faults and fractures (Fig. 4.6).

In addition to lunar sinuous rilles, many other surface features are recognized on maria. Prominent among these are various so-called wrinkle or mare ridges. Mare ridges may extend tens of kilometers across the mare surfaces and usually are composed of a broad, basal arch, surmounted by a sharply crenulated, sinuous element (Fig. 4.19). Furthermore, Lucchita (1976) has demonstrated that some mare ridges are directly related to highland scarps and are probably the result of vertical faulting, as shown in Figure 4.20.

The origin of mare ridges is enigmatic. On the one hand, there is evidence to show that they post-date the formation of the maria in which they occur—many of them intersect impact craters (Fig. 4.21), which would indicate that the mare crust had solidified sufficiently to preserve the crater. On the other hand, some segments show flow features that suggest lava was extruded from beneath the crust at the time the ridge was formed (Fig. 4.22), thus suggesting that the ridge formed relatively contemporaneously with the emplacement of the mare lavas. Figure 4.23 shows one possible explanation that accounts for both the preservation of craters and the extrusion of material from beneath the mare crust. Other ridges reflect underlying topography, such as buried crater rims (Fig. 4.24) and basin rings.

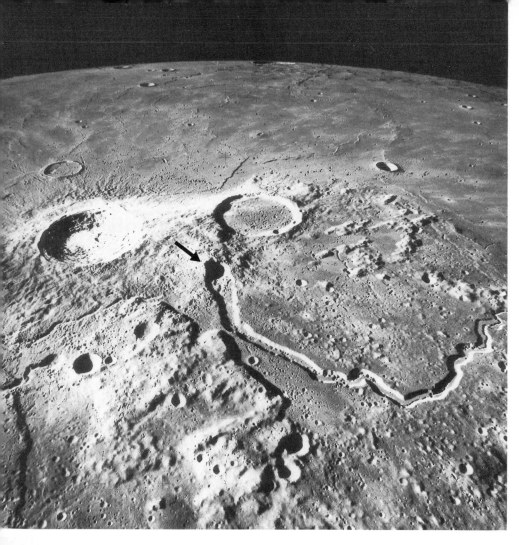

Figure 4.15 Oblique view taken by the Apollo 15 astronauts of the Aristarchus plateau and Schröter's Valley. The 40 km fresh impact crater in the upper left is Aristarchus. Schröter's Valley is more than 250 km long and exceeds 10 km in width. This rille originates in the "Cobra head" (arrow) and is considered to be a tectonic feature controlled by faulting combined with a volcanic flow channel which served as a vent for lavas emptying into Oceanus Procellarum to the upper right. The Aristarchus plateau is one of the major volcanic centers on the Moon. Illumination from the left (Apollo 15 AS15-M3-2611).

Figure 4.16 Lunar orbiter view of the Marius Hills, a volcanic center located in Oceanus Procellarum. Numerous domes, cones, and sinuous rilles occur in this area and are considered to be volcanic in origin; area shown is about 75 by 85 km (Lunar Orbiter V-M214).

Figure 4.17 Lunar orbiter view of the Rümker Hills, a volcanic platform 70 km across in northwest Oceanus Procellarum; numerous domes, some of which exceed 10 km across, are considered to be volcanic (Lunar Orbiter IV-H170).

Figure 4.18 Apollo metric camera view of the Hadley Rille region, area of the Apollo 15 landing site (A). This site, situated in the Montes Apennines (mountainous areas on the right half of the photograph, H) was selected to sample rim material of the Imbrium basin and to examine the mare lavas associated with the rille (AS15-414). ▷

Figure 4.19 Apollo metric camera view of lunar mare ridges in Mare Serenitatis showing the broad, gently swollen base surmounted by a sharply crenulated steeper portion. Note that some ridges cross from the younger light mare into the older (indicated by the higher frequency of craters), dark mare. Also shown are linear rilles, considered to be grabens. Area shown is about 150 by 170 km (AS17-0451).

Figure 4.20 Part of panoramic camera photograph of the Apollo 17 landing site (arrow) and a bright deposit overlying the dark mare plain. This deposit covers more than 21 km^2 and originated from the massif to the south (bottom of picture), and may be either an avalanche, or secondary crater and bright ray material. Shown also is the Lee-Lincoln scarp which cuts across the mare valley and continues as a fault onto the highland block to the north. Area shown is 24 by 22 km, illumination from the right, north to the top (AS17 M-1220).

Figure 4.21 Impact crater (6 km in diameter) in Mare Cognitum that has been cut by a fault-ridge segment (from Lucchitta 1976; Apollo 16 panoramic frame AS16-5429).

Figure 4.24 "Ghost" crater (about 20 km in diameter) near the Herigonius rille complex in southeastern Mare Cognitum showing part of the rim buried by mare lavas and exposed as mare ridges (from Young *et al.* 1973; AS16-M2834).

Figure 4.22 Apparent extrusion of lava from mare ridge superimposed on the crater wall (arrow) in eastern Mare Serenitatis, demonstrating the extrusive origin for parts of some mare ridges. Illumination from the left (from Hodges 1973; Apollo 15 panoramic frame AS15-9303).

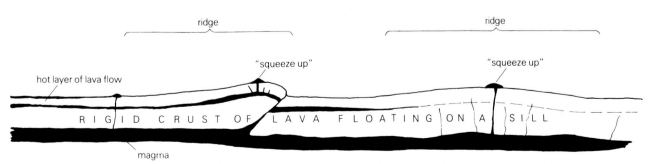

Figure 4.23 Diagram illustrating a hypothetical cross section through mare ridges of two different types. On the right is a ridge generated primarily by tectonic processes; on the left is an asymmetrical ridge resulting primarily from igneous processes (from Guest & Greeley 1977).

93

Figure 4.25(a) Oblique view of the crater Alphonsus showing prominent dark-haloed craters (arrows) considered to be pyroclastic vents (some of which are associated with fractures) in the floor of this ∼ 125 km impact crater. Device on the left side of photograph is the boom for the gamma-ray spectrometer carried on the spacecraft (AS16-M2478).

Figure 4.25(b) Lunar orbiter view of small structures in Oceanus Procellarum transecting the crater Hortensius D; these features are a few hundred meters across and are considered to be a row of spatter cones (Schultz 1976c; Lunar Orbiter IV-H133).

Figure 4.25(d) Small shield volcanoes about 10 km across in the Lacus Veris mare region of the Orientale basin. Illumination from the right (from Greeley 1976; Lunar Orbiter IV-H181).

Figure 4.25(c) Apollo 15 view of the so-called "Caldera D" structure (named Ina); this feature is unique on the lunar surface and its origin is enigmatic, although most investigators attribute its origin to volcanic processes (Whitaker 1972, Strain & El-Baz 1980; Apollo 15 panoramic frame AS15-10181).

Figure 4.26 Oblique Lunar orbiter view of the Alpine Valley. This 150 km long by 10 km wide graben is radial to the Imbrium basin and is considered to be a tectonic feature associated with the formation of the basin. The floor of the graben has been flooded with mare lavas which have been emplaced through the rille seen in the center of the valley. View toward the south (Lunar Orbiter IV-M102).

4.2.6 Mare domes and other small features

Numerous domes (Fig. 4.17) have been identified on the Moon. They range in size from <1 to >20 km across. Head and Gifford (1980) have mapped more than 200 mare domes and recognize two types. The first type consists of low, relatively flat-topped circular structures which often have summit craters. They consider this type to be volcanic domes formed as a result of relatively low rates of eruption. Morphologic analyses by Schultz (1976*c*) show that many domes are furrowed and display features that could represent flow of viscous lavas. The second type of mare dome is considered to be high-standing "islands" of terrain remaining unflooded by younger mare deposits.

In addition to dome volcanics, other possible volcanic vents have been identified. As shown in Figure 4.25, these include dark halo craters which may be pyroclastic features, small shield volcanoes and possible spatter cones.

4.3 Tectonic features

Most tectonic features on the Moon result from crustal deformation associated with impact basins, or with adjustments inferred to be associated with the cooling and settling of mare lavas. The Alpine Valley is a graben 10 km wide which cuts across the Alpes mountains for 150 km (Fig. 4.26).

Numerous **linear rilles** criss-crossing the surface of the Moon attest to the tectonic deformation of the lunar crust. Unlike sinuous rilles which are generally controlled by topography, these rilles have the general form of grabens (Golombek 1979) and appear to be relatively deep-seated structures which cross both highlands and mare units (Fig. 4.27). Sequence and timing of the deformation can often be demonstrated from superposition and cross-cutting relations, as shown in Figure 4.28, where a rille is seen to cut part of the Montes Apennines (part of the rim structure for the Imbrium basin) and has in turn been flooded by mare units. Thus, it is clear in this region that the crustal faulting occurred after the formation of the basin but prior to the emplacement of the surface lava.

Figure 4.27 Apollo 10 oblique view of Rima Ariadaeus near the center of the lunar nearside. This linear rille is about 2 km wide and cuts across both mare and highland topography indicating that it is a deep-seated tectonic feature, probably a graben (Apollo 10 frame 10-4646).

Figure 4.29

Figure 4.28 Apollo 15 view of linear rilles (grabens) southeast of Hadley Rille (the sinuous rille in the upper right); illumination from the right (Apollo 15 frame M-1820).

Figure 4.29 Floor-fractured craters bordering western Oceanus Procellarum; largest crater is about 75 km across (Lunar Orbiter IV-H189).

Figure 4.30 Diagrams showing the development of floor fractured craters. The dashed line in the top figure shows the initial excavation cavity before adjustment of the floor and wall collapse immediately following impact (from Schultz 1976a).

Impact

Floor fracturing

Floor uplift

Mare flooding

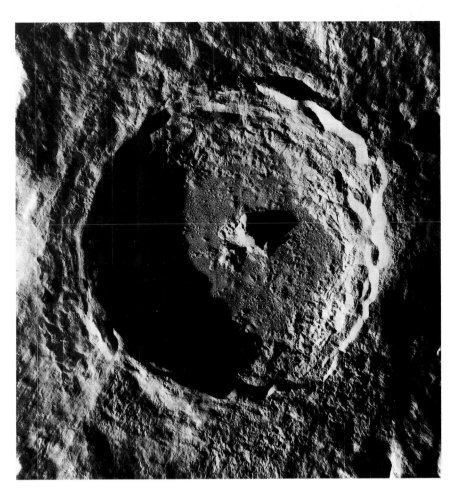

Figure 4.31 The crater Tycho (85 km in diameter) is an excellent example of a large, fresh, lunar crater. Typical of this crater class is a high central peak, slumped walls and hummocky rim deposits. The floor is extremely rough, with mounds and fissures. The rim displays prominent flow features and a strong concentric texture in many places; north is to the bottom of the photograph (Lunar Orbiter V-M125).

Study of the global patterns of lunar grabens and their timing of formation as related to the emplacement of lavas (Lucchitta and Watkins 1978) shows: (a) most grabens formed after the emplacement of the early-stage mare lavas and considerably after basin formation; (b) the grabens formed on faults that were reactivated along older structures; (c) large graben systems ceased activity at about 3.6 ± 0.2 eons; and (d) the grabens reflect a tensional stress field generated in the early history of the Moon.

The floors of many larger craters and small basins show evidence of tectonic deformation (Fig. 4.29). Termed by Schultz (1976*a*) **floor-fractured craters,** the deformation may have occurred as a consequence of magmatic intrusion and the emplacement of dikes, shown in Figure 4.30. Hall *et al.* (1981) argue on geophysical grounds that at least some floor-fractured craters could result from viscous relaxation of the crater topography.

4.4 Craters

Craters on the Moon can be classified as: (a) *impact basins* (Fig. 2.5), described above, (b) *large impact cra-* *ters*, those larger than about 15 km across (Fig. 4.31), (c) *small impact craters* (Fig. 4.32), which range in size from micrometers to those 15 km in diameter, and (d) *non-impact craters*. The transition from large craters to basins on the Moon is generally considered to occur in the size range of 140–220 km.

Studies by Hale and Grieve (1982) show that the morphologic and volumetric characteristics of central peaks change in the range of crater diameters of 51–80 km and grade into peak rings. Peak ring impact structures grade, as their diameter increases, into ringed and multi-ringed basins—suggesting to some researchers that craters and basins are not fundamentally different and that basins should not be treated as a special category but simply as very large craters. This concept is consistent with various theories of the origins of central peaks, peak rings and interior multi-rings.

The distinction between fresh, large craters and small craters can be seen in their cross-sectional geometry. For example, Pike (1980) and other workers have demonstrated a distinct break in the depth-to-diameter ratio, as shown in Figure 4.33. Generally, large craters are proportionally more shallow than small craters. The difference in the depth to diameter ratios between small and large craters (up to ~ 100 km) is explained by slumping

Figure 4.32 View of a typical small (8 km in diameter) bowl-shaped crater on the Moon (AS10-29-4253).

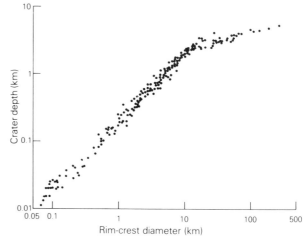

Figure 4.33 Depth to diameter ratios for 204 fresh lunar craters, as compiled from Apollo 15, 16 and 17 data. Distribution inflection at a diameter of about 15 km indicates the transition from small, simple craters to large, complex craters (from Pike 1974; *Geophys. Res. Let.,* copyright American Geophysical Union).

Figure 4.34 Oblique view of Necho crater on the lunar farside showing extensive wall slumping (AS10-28-4012).

and infill of large crater floors. In addition, large craters tend to have central peaks in a variety of arrangements (e.g. ridges and rings) and complex wall structures, including terraces formed by slumping (Fig. 4.34). In many cases the floors of large craters appear to have been modified by various post-impact processes which may include volcanic eruptions that could form domes and smooth floor deposits, as seen on the floor of Copernicus (Fig. 4.35). Crater floors are frequently covered with blocks, fissures and mounds. Long fissures commonly parallel the floor-wall contact, while shorter, more irregular ones have variable directions. Some mounds are surrounded by symmetrical rings or shells which most researchers think may be due to the flowing of shock-fluidized material off the surface of the mounds.

The zones outside the rims of large craters often show extensive modifications, including scouring and erosion by ejecta and ejecta-related melt deposits, some of which flowed down the flanks of the crater to form smooth "pools" where it ponded (Figs. 4.36 and 4.37). An apron-like expanse of striated material beyond the northwest rim of the prominent mare-filled farside crater Tsiolkovskiy is interpreted as a giant landslide (Fig. 4.38), similar to the Sherman landslide in Alaska (Fig. 3.31(d)) which was triggered by an earthquake. In contrast, small craters tend to be simple, bowl-shaped depressions, although modifications of the wall by slumping are also visible in some cases.

As discussed in Chapter 3, impact cratering involves the point-source transfer of energy from the bolide to the target surface. This typically results in craters and crater ejecta deposits that are radially symmetrical. However, for angles of impact lower than about 15° from the horizontal, both craters and ejecta deposits tend to become asymmetric with rays that frequently form a "wing" pattern, as seen by several examples on the Moon (Fig. 4.39).

Ejecta deposits for all fresh lunar craters larger than a kilometer can be divided into three relatively well-defined regions which are of increasing distance and usually (with the above-noted exceptions) concentric about the rim

Figure 4.35 Lunar orbiter view of the floor of Copernicus showing numerous domical structures (some of which have summit craters), fissures, and smooth floor deposits, all possibly indicative of magmatic activity; the largest domical feature is ~ 1 km across (Lunar Orbiter V-H154).

Figure 4.36 Oblique view of King crater (~ 75 km) on the lunar farside showing smooth deposits (arrow) considered to be melted ejecta associated with the impact that formed the crater. Similar flow deposits are observed around many craters of this size (AS16-MO488).

Figure 4.37 A closeup of the "pools" of material observed beyond the north rim of crater King in Figure 4.36. Cracks and channels in the pools and numerous small ridges suggest that the material was molten (AS16-5000).

crest. The inner zone is characterized by blocks and concentric fractures. Larger craters may exhibit "pools" of smooth-surfaced units and dune-like features. The middle zone is a region of continuous ejecta which exhibits a hummocky surface with radial and concentric patterns, some of which appear dune-like. The outermost zone is of discontinuous ejecta which is composed of filamentary rays and numerous secondary craters which frequently display a characteristic herringbone pattern. Large impact craters, such as Copernicus (Fig. 4.40), possess both concentric and radial patterns of secondary crater chains.

Some volcanic craters are morphologically similar to impact craters. Thus, while the vast majority of lunar craters are considered to be of impact origin, it is possible that volcanic craters are also present. Some linear arrays of craters almost certainly are of internal origin. Chains of craters of possible non-impact origin are classified (e.g. Eppler and Heiken 1975, Pike 1976) into: (a) collapsed lava tubes, (b) cinder-cone craters and (c) drainage pits. This latter category includes the Hyginus Rille (Fig. 4.41) which exhibits a series of rimless craters several kilometers across, all contained within a graben. These are generally considered to be collapse craters, possibly formed in association with magmatic withdrawal.

4.5 Degradational features

Because the Moon lacks an atmosphere and has never experienced liquid water on the surface, lunar degradational processes are restricted primarily to mass wasting and various impact-related events. Mass wasting has occurred in several forms, including possible landslides (Howard 1973; Fig. 4.20), rock-falls and slides (Fig. 4.42), and surface creep leading to **treebark textures.** Treebark texture, shown in Figure 4.43, is considered to result from small-scale impact cratering. Slumping of crater walls is particularly important in the degradation of craters (Fig. 4.36). Impact cratering is also responsible for the generation of lunar regolith deposits. Lunar surfaces are constantly being fragmented by impact cratering at all scales. This constant pulverization has led to the formation of regolith deposits, the thickness of which is directly proportional to the age of the surfaces. Thus, young mare lavas are blanketed with a fragmental surface layer estimated to be only a few meters thick, while older mare surfaces may have regolith deposits exceeding 10 m in thickness (Oberbeck and Quaide 1968). Figure 4.44 shows the lunar surface viewed by the Apollo astronauts

Figure 4.38 A large lobate sheet of striated material on the northwest flank of Tsiolkovskiy crater (visible on the right) interpreted as a giant landslide. The distance between the head scarp, just beyond the rim of the crater and the outer edge of the flow is ~ 80 km (AS 17 M-2608).

Figure 4.39 View of Messier (1) and Messier A (2) in Mare Fecunditatis; the shape of these craters and the asymmetric ejecta pattern are considered to be the result of a low-angle impact in which the bolide either broke into two components, or "skipped" across the surface to form two craters. Messier is about 8 by 15 km (AS15-M3-2405).

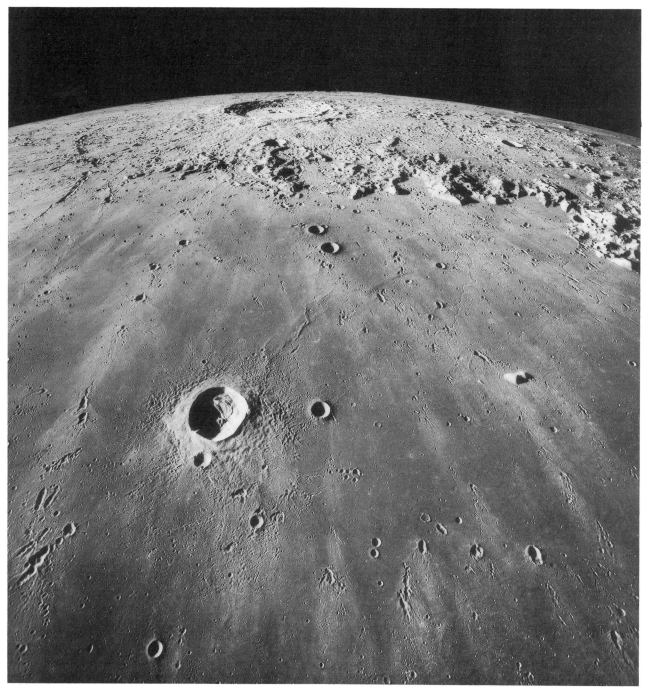

Figure 4.40 An oblique view across southern Mare Imbrium looking toward Copernicus, the large (93 km in diameter) crater near the horizon. The distance from the lower edge of the picture to the center of Copernicus is 400 km. The mountains at the edge of the mare are the Montes Carpatus, a major ring of the Imbrium basin. Copernicus is one of the youngest of the Moon's large craters. The many chains, loops, and clusters of small irregular craters and associated bright streaks extending across the mare are caused by the secondary impact of ejecta from Copernicus (AS17-M2444).

Figure 4.41 Lunar orbiter view of the Hyginus Rille. This rille extends for more than 200 km and contains a series of rimless craters (some as large as 12 km in diameter) considered to be similar to volcanic pit craters and ascribed to endogenic processes (Lunar Orbiter V-M97).

and gives a general indication of the formation of regolith through impact processes.

4.6 History of the Moon

Photogeologic mapping and remote sensing, analysis of various surface features and incorporation of results from lunar sample studies have yielded a general model for the evolution of the lunar surface. Figure 4.45 illustrates the major stages in the history of the Moon, keyed to the lunar stratigraphic timescale (Wilhelms 1980) in which systems recognized from oldest to youngest are: (1) pre-Nectarian, (2) Nectarian, (3) Imbrian, (4) Eratosthenian, (5) Copernican. The pre-Nectarian system represents all of the older cratered terrains which were formed prior to the impact of the Nectaris basin. This interval of time records the final stages of the accretionary process in the formation of the Moon. Impact-generated heat is considered responsible for the melting of the outer few hundred kilometers and the separation of an anorthositic crust. By about 3.9 eons, the outer crust was rigid and volcanism was deep-seated.

The Nectarian system begins with the formation of the Nectaris basin and the emplacement of its ejecta blanket, the Janssen Formation. This interval of time includes the formation of other older basins—including Humorum and Crisium—and eruptions of high-aluminum mare lavas and lavas rich in potassium (K), rare earth elements (REE), and phosphorus (P) collectively termed KREEP and KREEPy lavas. Studies by Hawke and Head (1978) show that KREEP volcanism began prior to 4.1 eons and continued into early to mid-Imbrian time.

The impact that formed the Imbrium basin (Fig. 4.45(a)) occurred about 3.85 eons and marked the beginning of the Imbrian system. This interval of time includes the formation of the Orientale basin and the eruption of about two-thirds the volume of the visible maria (Fig. 4.45(b)).

The Eratosthenian system is exemplified by impact craters similar to Eratosthenes which have well-preserved secondary craters but which lack bright-ray deposits. During this interval of time about one-third the remaining portion of the visible maria were emplaced. The genera-

Figure 4.42 View obtained by the Apollo astronauts using a telephoto lens to photograph boulders and boulder tracks on the slopes of North Massif at the Apollo 17 landing site. The largest boulders are about 5 m across; illumination is from the right (AS17-144-21991).

Figure 4.43 Lunar orbiter view of a small (~ 600 m) crater showing "treebark" texture on the crater wall; this pattern is considered to result from impact-generated creep (Lunar Orbiter II-H56).

tion of late-stage high-titanium mare lavas, plus some low-titanium, intermediate-stage lavas also occurred during this period.

The last episode of lunar history is identified as the Copernican system, which includes all fresh, large craters that retain bright rays. Photogeologic mapping of some presumed mare lava flows superimposed on Copernican-age craters showed that some lunar volcanism may have extended to relatively recent times (post ~ 2.0 eons) on the lunar surface.

Figure 4.44 View near the Hadley Rille in the Apollo 15 landing site area, showing typical block-littered surface in which the rocks are being physically weathered by impact cratering (AS15-82-11147).

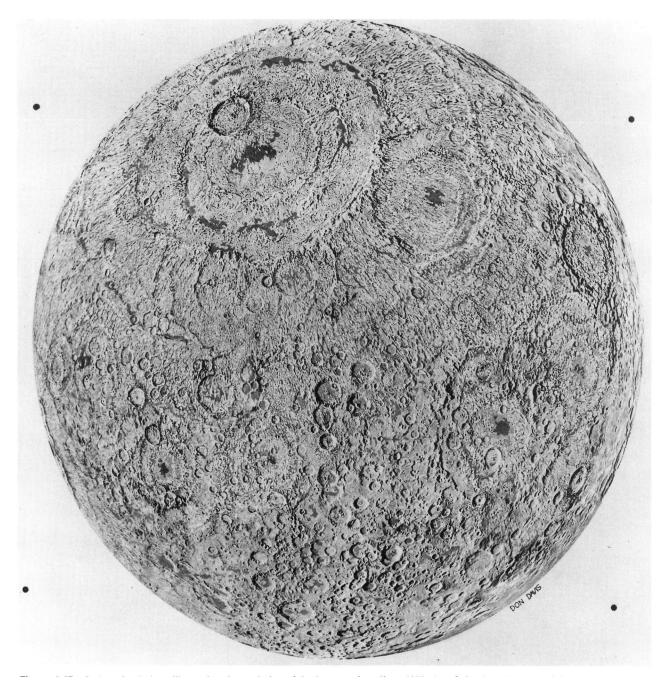

Figure 4.45 Series of paintings illustrating the evolution of the lunar surface (from Wilhelms & Davies 1971, copyright © 1971 Academic Press). **(a)** The Moon during the middle of the Imbrium period, after the formation of most of the large impact basins, but before extensive flooding by mare lavas; the Imbrium basin is in the upper left.

Figure 4.45(b)   The Moon after extensive eruptions of mare lavas at about 3.3 eons ago, but before the formation of craters such as Eratosthenes, Tycho and Copernicus.

Figure 4.45(c)   The present Moon.

5 Mercury

5.1 Introduction

Mercury, innermost planet of the Solar System, is easily visible to the naked eye and has figured prominently in folklore and mythology through the ages. Mercury derives its romanized name from the Greek god, Hermes, the Olympian patron of trade, travel and thieves. The first recorded observation of Mercury is by Timocharis, who noted its position in the heavens in 265 B.C. As viewed from Earth, Mercury is never more than 28° from the Sun and, consequently, it can be seen only in the twilight hours. Thus, even though it is an easy object to see, telescopic scientific observations are difficult. The Italian astronomer, Zupus, first recorded the phases of Mercury in 1639, while surface markings were not noted until 1800 when the astronomers Schröter and Harding reported albedo patterns.

Prior to acquisition of Mariner 10 results in 1974–5, relatively little was known of Mercury geologically. Estimates of its bulk density suggested that Mercury was

Figure 5.1(a)

Figure 5.1(b)

Figure 5.1(c)

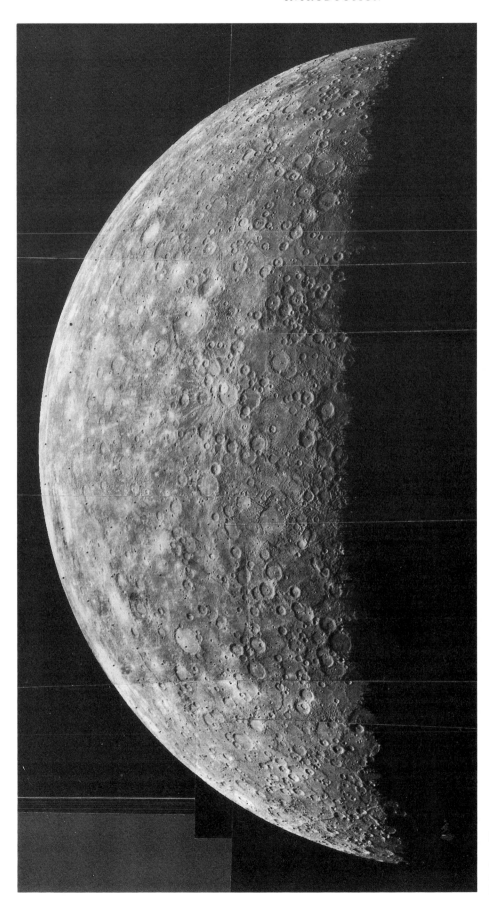

Figure 5.1 Mosaics of Mariner 10 images of Mercury obtained on the three flybys of Mariner 10 (courtesy NASA-Jet Propulsion Laboratory).

North polar region

South polar region

Figure 5.2 Shaded airbrush relief map of Mercury, showing prominent named features: (a, top) equatorial region, (b, left) polar areas (compare to Fig. 5.1). (Courtesy US Geological Survey, Flagstaff).

more like the Earth than the Moon (Table 2.1), which led to speculation that there could be twice as much iron in the core of Mercury in proportion to its size than any other planet in the Solar System. Earth-based observations of the reflective and thermal properties of the mercurian surface yielded results that were similar to those from the Moon's surface. This led to the suggestion that its surface is covered with a layer of dark, porous, fine-grained material which, as on the Moon, could have been generated by impact cratering processes.

It was not until radar observations were made in the early 1960s that the rotation period for Mercury was determined accurately. Mercury was found to have a two-thirds resonance with respect to its orbital period; that is, Mercury rotates on its axis precisely three times for every two orbits around the Sun. Various observations prior to the Mariner 10 mission also showed that Mercury has no significant atmosphere; thus, surface weathering or erosional processes involving running water or wind to modify its surface were not expected. Of course, it was not known if such processes might have occurred in the geologic past.

Mariner 10 opened Mercury to geologic inspection by providing the first (and as yet the only) detailed information for its surface (Murray *et al.* 1975, Strom *et al.* 1975*a*). Figure 5.1 shows a series of mosaics obtained on the three flybys of Mariner 10, the closest of which occurred at a distance of only 700 km. In all, approximately half of Mercury was covered and more than 2700 useful images were obtained (Murray 1975). Although these images have 5000 times better resolution than Earth-based telescopic views, it is important to keep in mind that most Mariner 10 images of Mercury are comparable to Earth-based telescopic views of the Moon (Table 2.3). On first inspection Mercury appears to be a Moon-like object (Fig. 5.2). In recognition of its similarity to the Moon, a major conference comparing the Moon and Mercury was held in 1976, with the proceedings published in a special issue of the journal, *Physics of Earth and Planetary Interiors* (1977; Table 1.2).

Like the Moon, the mercurian surface is dominated by impact craters. In addition, there is an asymmetry in the distribution of terrains: smooth plains dominate one part of the planet (Fig. 5.1(a)), whereas heavily cratered terrains (Fig. 5.1(b)) occur elsewhere, similar to the concentration of mare on the lunar nearside (at least for the half of Mercury for which data are available). Closer examination of the surface, however, reveals significant differences between Mercury and the Moon. First, there is no correlation between albedo and topography. The albedo of both the smooth plains and the cratered terrain is remarkably similar on Mercury, in contrast to the Moon with its dark-toned mare lowlands and light-toned highlands. An extensive terrain, termed **inter-crater plains,**

some of which may represent primordial crust, has been little modified throughout the evolution of the mercurian surface and has no direct lunar counterpart. Some aspects of the morphology of impact craters on Mercury are different in comparison to lunar craters. Finally, numerous scarps cut across plains, craters and cratered terrain on Mercury and suggest compressional crustal deformation.

In addition to the return of images, Mariner 10 made other measurements important to the interpretation of the geologic history of Mercury. It was found that the planet has a magnetic field, although it is very weak in comparison to Earth: 600 gammas versus 50,000 gammas for Earth. The origin of this magnetic field remains an enigma. It could be a remnant from some event in the past, a weak active field generated by magnetohydrodynamic processes, or a combination remnant/active field. In any event, the presence of the field, when combined with models of the interior that include a high proportion of iron and the suggestion of volcanism inferred from Mariner 10 images, suggests that Mercury underwent chemical differentiation.

In addition to the reference given in Table 1.2, several important reviews of Mercury have been published, including Guest and O'Donnell (1977) and Gault *et al.* (1977). Robert Strom's review (1979) is very extensive and is especially recommended. Davies *et al.* (1978) provide a synopsis of the history of studies of Mercury, a brief outline of its geology and cartography, and a superb atlas of photographs and charts keyed to the US Geological Survey system of quadrangles.

5.2 Physiography

Mariner 10 obtained images for only about half of the planet and thus the analysis of Mercury cannot be applied globally. Early in the analysis of the images, a geologic terrain map was derived and the major physiographic features were defined (Trask and Guest 1975). Following this preliminary mapping, the US Geological Survey embarked on a geologic mapping program based on 1:5,000,000 charts. While this mapping has allowed a refinement of the units and has filled in detail, the major units and the basic ideas regarding their origin have not changed significantly. A synthesis of all mapping is currently underway by the US Geological Survey which will lead to publication of a planet-wide geologic map in the mid 1990s.

Figure 5.3 is taken from Trask and Guest (1975) and shows the major terrain types and features on Mercury. Four major terrains are recognized: smooth plains, intercrater plains, heavily cratered terrain, and hilly and lineated terrain.

Key

	Rayed and dark-halo craters		Fresh craters and basins		Plains		Caloris mountains		Caloris lineated terrain		Hummocky plains	
	Hilly and lineated terrain		Hilly terrain		Lineated terrain		Heavily-cratered terrain		Old craters and basins		Craters cut by Caloris lineaments	
	Intercrater plains		Scarp: bar and ball on down side		Ridge		Fracture					

Figure 5.3 Map showing the principal terrain units on Mercury, based on the mosaics shown in Figure 5.1a & b (from Trask & Guest 1975, *J. Geophys. Res.,* copyright American Geophysical Union).

Figure 5.4 Smooth plains southeast of the Caloris basin showing a mare-like ridge; area shown is about 100 km across (Mariner 10 FDS 70).

5.2.1 Smooth plains

Smooth plains form relatively flat, sparsely cratered surfaces which resemble lunar maria and display bright-halo, sharp-rimmed craters generally smaller than ~ 10 km. Many smooth plains have sinuous, lobate ridges which are similar to lunar mare ridges but lack the upper crenulated ridge (Fig. 5.4). The plains often fill large craters and embay older terrains (Fig. 5.5) with well-defined contacts. From these contact relations and their lack of large, superposed craters, the smooth plains are considered to be among the youngest stratigraphic units on Mercury.

Smooth plains are found predominantly in the northern hemisphere between 120° to 190° longitude, where they form an approximate annulus around the Caloris basin (Fig. 2.18). However, smooth plains also occur as isolated patches elsewhere on Mercury, including the floors of large craters and basins. Regardless of location, smooth plains have about the same frequency of superposed impact craters, which suggest that the plains were formed at about the same time. Their spectral properties (color) are approximately the same at all locations, which suggests a homogeneous composition, unlike lunar maria or the Cayley plains. Moreover, smooth plains have about the same albedo as the heavily cratered terrain and intercrater plains in which they occur.

The origin of smooth plains is a matter of debate. Most of the controversy concerns the smooth plains found in the vicinity of Caloris. Based on the characteristics listed above, Wilhelms (1976) and Oberbeck *et al.* (1977) noted that the smooth plains east of Caloris are similar to the Cayley Formation and other lunar highlands and suggested that the mercurian smooth plains are also impact related. The difficulties with this interpretation are the apparent volume of material involved and the timing. From crater frequency distributions, there appears to be a substantial interval between the formation of the Caloris basin and the emplacement of the plains. Thus, it would seem unlikely that the plains could be either impact melt or ejecta-related deposits of Caloris.

The same difficulty applies to the study of smooth plains in other areas. Figure 5.5 shows Borealis Planitia, a circular depression (basin?) some 1000 km across, within which occurs Goethe, a 340 km impact basin. The plains flooding the depression and Goethe have a crater density significantly lower than the ejecta units from these large impact structures, indicating that the smooth plains post-date both the circular depression and Goethe and are unlikely to be associated with either impact event. However, it has been argued that these plains are related not to the Borealis basin or to Goethe, but to some unknown basin or basins hidden from view in the part of Mercury not imaged by Mariner 10.

Most investigators consider the smooth plains to be volcanic. Although there is no direct evidence, the following reasons are rallied for a volcanic origin.

1. Smooth plains are widespread over Mercury and are unlikely to have been the result of a single impact event; moreover, the smooth plains near Caloris appear to be of too large a volume to be solely Caloris ejecta.

2. The morphology and inferred physical properties of craters in smooth plains suggest similarities to lunar mare (Cintala *et al.* 1977). The presence of lunar-like mare ridges also suggests a similarity; because lunar mare are volcanic, the analogy is drawn for Mercury.

116

Figure 5.5 Photomosaic of Mariner 10 images in the north polar region of Borealis Planitia showing the Goethe basin (340 km diameter) and a ~ 1000 km unnamed basin (arrows mark rim), both of which have been flooded by smooth plains. Frequency distributions of superposed craters show that the plains are much younger than either Goethe or the large basin and hence are not related to the impacts which formed them (from Strom 1979).

3. Stratigraphic relations indicate that the smooth plains are much younger than the impact craters and basins with which they are associated.

5.2.2 Intercrater plains

The intercrater plains, shown in Figure 5.6, are the most widespread unit mapped and constitute about one-third of the area imaged by Mariner 10. This terrain consists of level to gently rolling plains peppered with craters smaller than 5–10 km, many of which appear to be secondary craters. The density of superposed craters is much higher than that of smooth plains. Preliminary mapping suggested that most intercrater plains were overlain by the heavily cratered terrain units and large, old craters.

Consequently, intercrater plains were interpreted to be the oldest unit on Mercury, pre-dating the period of heavy impact bombardment and perhaps representing relatively undisturbed primordial crust. Malin (1976) and Strom (1979), however, have discovered that some heavily cratered terrain is embayed (i.e. superposed) by intercrater plains (Fig. 5.7). Furthermore, Spudis and Strobell (1984) find a large population of ancient multi-ring basins which underlie the intercrater plains on a planet-wide basis, suggesting that the plains are not primordial crust.

It is important to note that the intercrater plains should not be treated as a stratigraphic unit (Guest and O'Donnell 1977); this terrain unit probably represents different events separated not only geographically but by time. Leake (1981) analyzed the intercrater plains and relative ages of cratered terrain and individual large, old craters

Figure 5.6 High oblique view showing large areas of intercrater plains and heavily cratered terrain. The apparent "tear" toward the top of the picture was caused by a loss of image data and is an artifact. This image was taken at a distance of 77,800 km from Mercury (Mariner 10 FDS 27328).

Figure 5.7 High-resolution Mariner 10 image showing typical intercrater plains and the superposition of numerous craters; arrow points to a 90 km diameter crater that has been embayed by intercrater plains and, thus, pre-dates the emplacement of the plains. The lobate scarp, Santa Maria Rupes, post-dates the intercrater plains. Image is 400 km wide (from Strom 1979; Mariner 10 FDS 27448).

and concluded that at least some intercrater plains were being formed concurrently with heavy impact cratering. Thomas (1980) and Thomas *et al.* (1982) studied the relationships between the plains and nearby impact craters as a function of their stage of degradation and concluded that intercrater plains formation has occurred throughout the history of Mercury. Strom (1979) carried this possibility one step further and concluded that the intercrater plains and the very young smooth plains may represent a continuous process of plains formation, beginning with the earliest observable history of Mercury and extending through to the latest events.

The origin of the intercrater plains, like that of smooth plains, remains an enigma. Stratigraphic relationships led to the conclusion that intercrater plains are very old, perhaps representing primordial crust. However, recognition that intercrater plains span a wide range of ages led to the idea that the plains are not a single unit and may not have a single mode of origin.

Like the mercurian smooth plains, the resemblance of intercrater plains to lunar maria suggests a volcanic origin. Some of the intermediate-age plains fill large degraded basins and, based upon the appearance of ghost craters in these plains, may be relatively thin units (De-Hon 1978), such as thin flood lavas.

The intercrater plains have been likened to the pre-Imbrian pitted plains as mapped by Wilhelms and McCauley (1971) on the Moon and, like the lunar plains, alternative origins to volcanism have been proposed for the mercurian plains. Wilhelms (1976) and Oberbeck *et al.* (1977) suggest that the plains are generated by impact cratering processes and could be ejecta deposits. Although this origin could explain the interleaving of intercrater plains with the formation of large craters, the difficulty with this interpretation is the lack of source basins for the ejecta.

5.2.3 *Heavily cratered terrain*

Heavily cratered terrain consists of closely packed and overlapping large craters (Fig. 5.6), along with individual craters ranging in size from 30 km to several hundred kilometers across. In most of the heavily cratered terrains, clearly defined ejecta deposits and fields of secondary craters are not identified, although some of the craters on the intercrater plains may be secondaries from craters in this unit (Trask and Guest 1975). Many of the large craters in this terrain are filled with smooth, much younger, sparsely cratered plains (Fig. 5.8). The similarity between the heavily cratered terrain and the lunar highlands leads most investigators to consider this unit to be the consequence of the terminal phase of heavy bombardment which evidently occurred throughout the inner Solar System (Gault *et al.* 1977).

Figure 5.8 Image of the south polar area centered at 69°S, 104°W, showing Bach, a ~ 200 km diameter ringed basin filled with smooth plains (Mariner 10 mosaic).

5.2.4 *Hilly and lineated terrain*

Hilly and lineated terrain consists of an unusual landscape covering more than 250,000 km², centered at about 20°S, 20°W. Informally termed the "weird terrain" by the Mariner 10 imaging team, this landscape consists of hills 5–10 km across and 0.1–1.8 km high. Most of the crater rims in this area have been broken into numerous massifs, hills and depressions as shown in Figures 5.9 and 5.10.

The hilly and lineated terrain is antipodal to the Caloris basin, a location which suggested to Schultz and Gault (1975) that it may be related in some way to the Caloris impact. They analyzed the partitioning of seismic energy that would be related to an impact of the magnitude required to form the Caloris basin and predicted that there would be a tremendous focusing of energy at the antipode (Fig. 5.11), resulting primarily from the presence of a large planetary core. Schultz and Gault estimate that the mercurian surface would have been lifted tens of meters by seismic waves at the antipode which could account

Figure 5.9 Hilly and lineated ("weird") terrain centered at about 27°S, 24°W antipodal to the Caloris basin. Crater rims and intercrater areas have been broken into hills and valleys, whereas most of the smooth plains filling Petrarch (the ~ 150 km crater on the right) have not been modified. This suggests a sequence beginning with cratering, followed by terrain modification, and ending with the emplacement of smooth plains. The episode of terrain modification correlates with the formation of the Caloris basin. Arrow points to crater shown in Figure 5.10 (from Strom 1979; Mariner 10 FDS 27370).

for the broken and "jostled" appearance of the hilly and lineated terrain.

5.3 Craters and basins

Impact craters, basins and related structures are the dominant landforms on Mercury. In this section the morphology of impact craters and basins is discussed and compared to lunar craters and the results from crater counts are assessed to determine the possible sequence of events in the evolution of Mercury.

5.3.1 Morphology

Initial examination of mercurian impact craters suggests a similarity to craters on the Moon. This is not surprising since both objects are airless bodies. Closer scrutiny, however, shows significant differences between mercurian and lunar craters, especially in regard to rays, ejecta and secondary craters. Bright rays are more prominent (Fig. 5.1(a)), the continuous ejecta facies seems closer to the crater rim and secondary craters appear to be deeper, have sharper rims and occur closer to the parent crater rim than on the Moon (Fig. 5.12). Gault *et al.* (1975) attribute some of these differences to the higher gravitational acceleration on Mercury by arguing that

Figure 5.10 High-resolution image of hilly and lineated terrain; prominent crater on the left is about 40 km across and is identified in Figure 5.9 (Mariner 10 FDS 27463).

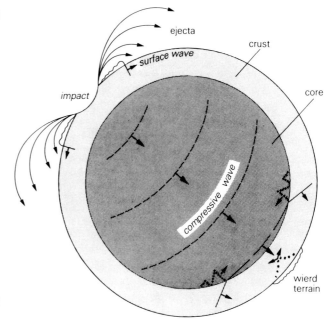

Figure 5.11 Diagram showing partitioning of energy resulting from major impact on a differentiated object such as Mercury; surface waves and compressive waves focus on the antipode and can result in severe seismic deformation (from data of Schultz and Gault 1975; courtesy of P. Schultz, Lunar and Planetary Institute).

121

Figure 5.12 Mariner 10 view of a 140 km diameter crater in the Shakespeare region showing the narrow hummocky rim, radial grooves and ridges, and field of secondary craters typical of large mercurian craters (Mariner 10 FDS 166).

Figure 5.13 Average radial extents of continuous ejecta deposits around fresh craters on Mercury and the Moon (after Gault *et al.* 1975; *J. Geophys. Res.*, copyright American Geophysical Union).

ejecta would travel a shorter distance there in comparison to the Moon (Fig. 3.6). They show that ejecta should travel only about 0.65 as far on Mercury as on the Moon (Fig. 5.13). Thus, the gravitational effects would cause the ejecta to form a much thicker deposit near the rim of mercurian craters than on the Moon.

Scott (1977) also considered secondary impact cratering velocities and found that due to the higher gravity on Mercury, velocities would be about 1.5 times greater than on the Moon. He attributed the sharper rims and apparently deeper secondary craters on Mercury to these more energetic impacts.

Mercurian craters display size-dependent morphologies that begin with small, bowl-shaped craters and progress to larger craters that have flat floors, central peaks, wall terraces (Fig. 5.14) and more complicated ejecta

deposits. At approximately 140 km, most mercurian craters begin to display concentric rings (Fig. 5.8) and the transition to basins occurs. With a diameter of 1300 km, the Caloris basin (Fig. 2.18) is the largest known basin on Mercury.

The depth-to-diameter ratio for mercurian craters is similar to that for lunar craters except that the transition to more shallow forms for large craters occurs at a diameter of 7–8 km rather than ~ 14 km, as on the Moon (Fig. 4.33). Although Gault *et al.* (1975) attribute this to the higher gravity on Mercury, Malin and Dzurisin (1977; 1978) find that the difference in the depth-to-diameter ratio between mercurian and lunar craters is small and consider gravitational effects to be minimal.

Several investigators have examined the morphology of crater interior features, such as central peaks, of the mercurian craters and attempted to correlate the results with terrain types for comparison with the Moon. Malin and Dzurisin (1977; 1978) concluded that the onset of central peaks occurs at about the same diameter on Mercury as on the Moon and that their formation appears to be independent of the terrain type. Cintala and co-workers (1977) conducted a more detailed analysis and arrived at the same conclusion. In addition, they found that the morphologies of craters formed in mercurian smooth plains and heavily cratered terrain is essentially the same as the morphology of craters in lunar maria. However, they found a greater difference in crater morphology for the lunar highlands in comparison to mercurian cratered terrain, leading them to conclude: (a) that differences

Figure 5.14 Oblique view of a 98 km diameter crater in the Shakespeare region, showing well-developed wall terraces and central peak (from Gault *et al.* 1975; Mariner 10 FDS 80).

in gravity have not dominated the formation of crater morphologic features, such as central peaks; (b) that the lunar mare/mercurian smooth plains similarities suggest that the smooth plains are volcanic, or at least have similar physical properties to the lunar maria; and (c) that morphologic differences in craters found in the heavily cratered terrain on Mercury and lunar highlands imply that these two terrains are dissimilar in their physical properties.

5.3.2 Crater states of preservation

At least four processes degrade craters on Mercury: (a) erosion from impact processes, including ballistic sedimentation; (b) burial or modification by plains units, such as lava flows; (c) isostatic adjustment (this process is particularly appropriate for the larger craters and basins); and (d) seismic events which may induce wall failures and crater rim degradation.

Several schemes to classify craters by their state of preservation have been devised. Most of these schemes are similar to those proposed for the Moon and are based on factors such as presence or absence of bright rays, preservation of secondary craters and sharpness of the rim. Classification involving three or more classes has been proposed but, unfortunately, the class numbering schemes have not been consistent. In some schemes, Class 1 refers to the freshest (i.e. youngest) crater, while in other schemes, Class 1 refers to the most degraded craters. The US Geological Survey mapping program utilizes a five-fold scheme with Class 5 representing the freshest craters.

Schultz (1977) examined mercurian craters and noted

123

Figure 5.15 Mariner 10 image of the 120 km diameter crater Zeami (3.1°S, 147.2°W), showing smooth dark plains and bright patches on the floor. The shallow floor, the possible rimmed moat and the floor deposits are similar to volcanically modified lunar craters; arrows indicate outer boundary of dark halo (from Schultz 1977).

Figure 5.16 Mariner 10 image showing well-preserved secondary crater field surrounding an old, degraded 190 km crater, Ma Chih-Yuan, the Discovery Quadrangle; arrows mark crater rim (from Schaber *et al.* 1977).

the similarity of some modified craters to the floor-fractured craters on the Moon. As with the lunar forms, he considers these to be impact craters that have been modified by volcanic or magmatic processes. The evidence for internally modified craters includes: (a) presence of very shallow floors with floor fractures which may indicate uplift and suggest intrusion; (b) smooth, sparsely cratered plains on the crater floors which indicate emplacement long after the impact event and suggest volcanic extrusion (Fig. 5.15); and (c) the presence of dark-halo craters and dark deposits associated with the floor units of craters, similar to dark-halo craters observed on the Moon (Fig. 4.25(a)) which are considered to be volcanic. Schultz also noticed that there are spectral contrasts between the floor units and the surrounding terrain, which suggested to him differences in composition which could not be attributed to impact events but rather suggest volcanic processes.

5.3.3 Crater frequency distributions

Guest and Gault (1976) recognized three classes of craters based on stages of degradation and analyzed their frequency distribution. They found a general deficiency of craters ~ 50 km in diameter which they attributed to: (a) a change with time in the distribution of bolides which could have caused a lack of crater formation in certain size ranges; (b) during one stage of crustal evolution, the mercurian crust may have been semi-molten and incapable of preserving craters above a certain diameter range; and/or (c) an erosional process occurred which preferentially obliterated some craters of certain size ranges. Their analysis also showed that fresh craters smaller than 20–30 km appear to be lacking in the cratered terrain surrounding the Caloris basin. They speculated that the Caloris event (or impacts which formed other "hidden" basins) might have caused widespread seismic obliteration of the population of craters in question. Schaber and his colleagues (1977) noted that Ma Chih-Yuan is one of many basins/large craters on Mercury which appear to be in an advanced state of isostatic adjustment but have well-preserved secondary craters (Fig. 5.16). They cite these as evidence for impacts into a thermally active crust of relatively low viscosity.

Initial analysis of the Mariner 10 data suggested that there were few basins 300–1000 km in diameter on Mercury (Schaber *et al.* 1977; Frey and Lowry 1979). Subsequent studies by DeHon (1978) and Spudis and Strobell (1984), however, have revealed a number of basins in this size range that have been partly buried or modified by various plains units and hence are difficult to see. There may not, therefore, be a significant lack of basins on Mercury.

5.4 Scarps and ridges

One of the most important discoveries on Mercury is the presence of scarps and ridges apparently formed by compressional tectonism (Fig. 5.17). Classifying the ridges and scarps and mapping their distribution and orientation is important in understanding the evolution of Mercury because they provide clues to styles and timing of tectonic processes. The most extensive study of these features is provided by Dzurisin (1978) who classified their general morphology (scarps, ridges or troughs) and planimetric form (arcuate, lobate, irregular or linear).

5.4.1 Arcuate scarps

Arcuate scarps form curved cliffs that may be locally sinuous or irregular in outline. They range in length from 100 to 600 km and may be as high as 1.1 km. Arcuate scarps cut across both intercrater plains and large craters, and post-date most other tectonic features. Hero Rupes

Figure 5.17 (top right) Photomosaic of Discovery Rupes centered at 56.3°S, 38.3°W. The scarp is more than 500 km long and is about 2 km high, making it one of the largest scarps on Mercury. Two craters (a) 55 km diameter Rameau and (b) 35 km unnamed crater are cut by the scarp, suggesting thrust-faulting; arrow points to dome shown in Figure 5.30 (from Strom 1979).

Figure 5.18 Oblique view of Hero Rupes; this typical arcuate scarp may outline an old basin and crosses smooth plains and transects several large craters. Area shown is nearly 700 km wide (from Dzurisin 1978; Mariner 10 FDS 166842).

125

Figure 5.19 View of Vostock Rupes cutting ~ 75 km diameter Guido D'Arezzo crater; foreshortening of the rim (arrow) suggests 5–7 km of thrust-fault displacement and evidence of compression of the mercurian crust (Mariner 10 FDS 27380).

are tectonic features which resulted from compressional deformation. From the magnitude and timing of deformation, most investigators infer widespread compression of the crust early in the history of Mercury. Strom *et al.* (1975*b*) suggest that this compression was the result of cooling and shrinkage of Mercury's large iron core and calculate that a 1–2 km decrease in planetary radius could account for the observed crustal foreshortening. An independent analysis performed by Solomon (1976), based on models of lithospheric cooling, leads to a similar amount of crustal foreshortening. In both models the pattern of crustal deformation should be spherically symmetrical, which would generate randomly oriented compressional features. The random orientation assumes that there were no pre-existing fractures.

Melosh (1977) considered the consequences of tidal despinning for Mercury which might have been generated by solar tidal torques. He speculated that compressional stresses would have been oriented E–W, with NW–NE linear shear and N–S thrusting in the crust. Unfortunately, analysis of scarp orientations is not conclusive in testing these models and is further hampered by the lack of global image coverage for Mercury. Most investigators conclude that the arcuate scarps are tectonic features which resulted from global compressive stresses generated by a combination of tidal despinning and planetary cooling.

5.4.2 Lobate scarps

Lobate scarps form the margins of elevated lobes or "tongues" of terrain and have fronts that are 500–700 km long and stand 0.5–1.0 km high. Like arcuate scarps, lobate scarps transect intercrater plains and large craters (Fig. 5.20), which led Dzurisin (1978) to suggest that they result from tectonic uplift. He noted that linear segments of some lobes are parallel to nearby lineaments and inferred that their orientation might have been controlled by some pre-existing structural "fabric" such as fractures from large basins or from stresses generated by planetary cooling or tidal spindown.

Other investigators have suggested that lobate scarps are widespread lava flows. However, the lobes are so large and have such high fronts (implying flow thicknesses of up to 1 km), it is difficult to envision how such a viscous flow could be emplaced.

5.4.3 Irregular intracrater scarps

These scarps occur on smooth, flat crater floors and may be 100 km long and 400 m high. They exhibit no consistent planimetric form, some being linear, others being sinuous. As shown in Figure 5.21, some intracrater scarps cut crater walls, whereas others appear to be diverted by

(Fig. 5.18), a typical arcuate scarp, has a radius of curvature of about 370 km, is more than 500 km long and stands 0.7–1.0 km high. From cross-cutting relations with large craters, its formation appears to be contemporaneous with the period of heavy impact cratering.

Vostock is one of the most important arcuate scarps on Mercury. As shown in Figure 5.19, this scarp crosscuts a 75 km diameter crater, shortening the crater circumference and suggesting 5–7 km of thrust-fault movement; there is no evidence of strike-slip faulting along this scarp. Analysis of the geometry implies a fault-plane dip of about 11°.

Study of all mercurian arcuate scarps suggests that they

Figure 5.20 Lobate scarp (arrows) which cuts intercrater plains for ~ 700 km; Dzurisin considers this and other lobate scarps to be tectonically uplifted blocks of crust; other investigators suggest that lobate scarps mark the flow fronts of viscous lava flows (from Dzurisin 1978; Mariner 10 FDS 166591).

topography such as crater central peaks. In some cases, the scarps separate differences in floor textures and they appear to represent a contact of some sort; Strom and colleagues (1975*b*) suggest that these scarps may represent viscous lava flows. Alternatively, intracrater scarps may have resulted from local tectonic processes associated with isostatic adjustments of the crater floor, or from magmatic intrusion.

5.4.4 Linear ridges

Linear ridges are straight, positive relief features 50–300 km long and 0.1–1 km high. Those found in the intercrater plains tend to have rounded profiles and may transform into linear scarps. Linear ridges in the Caloris area which are radial to the basin typically are broad and have flat

crests. Figure 5.22 shows Mirni Rupes, which has been analyzed by Strom (1979) who suggests that it is actually a ridge (in contrast to "rupes" which designates scarps) which pre-dates the plains in which it occurs. His proposition is based on crater counts, and he suggests that the ridge may have formed early in the history of Mercury and then was later flooded by plains-forming materials.

5.4.5 Linear troughs

Linear troughs occur in the hilly and lineated terrain. They are 40–150 km long, 5–15 km wide, 100–500 m deep and are generally open-ended with the valley floor merging onto the surrounding terrain. Figure 5.23 shows Arecibo Valles and the sculpturing of the walls which may represent slumping and mass wasting. Strom *et al.*

Figure 5.21 Slightly oblique Mariner 10 image showing three intracrater scarps; scarp at (A) is directed around central peak of a ~ 80 km diameter crater; scarp (B) cuts smooth plains in a 90 km diameter crater; scarp (C) is atypical of this class in that it extends across the crater rim into the intercrater plains to the north; also, the features consist of two almost parallel scarps and resemble an uplifted block or horst (modified from Dzurisin 1978; Mariner 10 FDS 27428).

Figure 5.22 Oblique view of Mirni Rupes, a ridge in the Discovery Quadrangle at 37°S, 45°W, showing the broad, rounded cross section typical of linear ridges. Area shown is about 300 km across (Mariner 10 FDS 27420).

(1975b) suggested that linear troughs result from tensional forces and, therefore, are grabens.

5.5 The Caloris basin

Caloris (Fig. 2.18) is the largest structure observed on Mercury. Named because it lies near the subsolar point during Mercury's closest approach to the Sun, the Caloris basin is more than 1300 km across and bears many similarities to the lunar Imbrium basin. Ejecta and ejecta-related deposits dominate much of the terrain observed during the flyby of Mariner 10 (Fig. 5.1(b)). The formation of the Caloris basin marks an important time-stratigraphic event in the history of Mercury and its ejecta blanket is a critical datum plane, similar to the Fra Mauro Formation on the Moon.

Ejecta deposits extend outward from the Caloris basin as far as one basin diameter (Fig. 5.24). On the basis of geologic mapping of the Shakespeare and Tolstoj quadrangles, McCauley and his colleagues (1981) derived a formal rock-stratigraphic sequence and proposed the name Caloris Group for the mappable units associated with the basin.

The rim of the basin is defined by the Caloris Mountains, composed of the Caloris Montes Formation which consists of rectilinear massifs as much as 30–50 km across (Fig. 5.25) standing as high as several kilometers above the surrounding terrain. The mountains are probably up-

Figure 5.23 High-resolution image of Arecibo Valles in the hilly and lineated terrain, centered at about 26°S, 28°W. The walls of the ~ 13 km wide valley are scalloped and may have been modified by slumping and other mass-wasting processes (Mariner 10 FDS 27470).

Figure 5.24 Mosaic of Mariner 10 images showing the northeast part of the Caloris basin and the Shakespeare basin.

lifted blocks of pre-basin crust which are veneered by late-stage ejecta from Caloris. A weakly defined outer scarp lies 100–160 km beyond the Caloris Mountains northeast of the basin and forms the outer margin of the structure.

The Nervo Formation consists of patches of hummocky to rolling plains that occur in depressions within Caloris Montes and in the zone between the mountains and the outlying scarp. In some places it appears to be draped over older terrain. This unit is interpreted as fallback ejecta which may include some impact melt.

The Odin Formation is an extensive unit which lies beyond the Caloris Mountains and was mapped by Trask and Guest (1975) as consisting of hummocky plains and low, closely spaced to scattered hills 0.3–1 km across. In some places, the hills are aligned concentrically with the basin rim and give a corrugated appearance. This unit can be traced as far as 1100 km from the basin rim and constitutes one of the major ejecta deposits of Caloris.

The Van Eyck Formation is similar to the lunar Imbrium basin Fra Mauro Formation and has a lineated facies and a secondary crater facies. The lineated facies extends as far as 1000 km from the Caloris Mountains and consists of long, hilly ridges and grooves radial to the basin (Fig. 5.26). The Van Eyck Formation is considered to represent the outermost ejecta facies associated with the Caloris basin.

The Caloris basin is floored with smooth plains that morphologically resemble those outside the basin, except that they have been severely modified by tectonic processes that have generated sets of ridges and grooves (Fig. 5.27). The plains are not considered to be impact ejecta. Because of their similarity to smooth plains exhibited elsewhere on Mercury and their resemblance to lunar mare plains, the Caloris interior smooth plains are considered to be lava flows and the interpretation is supported by the presence of possible lava channels (Fig. 5.28).

The ridges within the Caloris basin have rounded crests and are 50–300 km long, 1–12 km wide, and 100–500 m high. They form roughly polygonal patterns not seen in other mercurian basins or on the Moon. The ridges tend to be concentric with the basin rim, radial with respect

Figure 5.25 Southeast side of Caloris basin showing the type locality for the Caloris Montes Formation, the uplifted crustal blocks which form the main ring of the Caloris basin. Image width is about 730 km (Mariner 10 FDS 229).

Figure 5.26 Northeast side of Caloris basin showing the radial lineated facies of the Van Eyck Formation beyond the Caloris Montes. Image width is about 590 km (Mariner 10 FDS 193).

Figure 5.27 Southeast side of Caloris basin showing the ridged and grooved smooth plains of the interior of the basin. Image width is about 275 km (Mariner 10 FDS 110).

to the basin center, and more numerous and larger toward the basin center. The troughs are 1–10 km wide, 100–500 m deep, and generally have flat floors and steep walls and form polygonal patterns. The troughs tend to be concentric with the rim and slightly radial to the basin center. Although some troughs transect the ridges, ridges are never seen to transect troughs (Fig. 5.29).

Based on geometry and transection relations, Strom *et al*. (1975*b*) suggest that the ridges formed as a result of compression as the floor of the Caloris basin subsided. Dzurisin attributes this subsistence to magma withdrawal beneath Caloris as lavas were extruded outside the basin to form the extensive smooth plains. Floor subsidence was followed by uplift of the basin center—perhaps in response to isostatic adjustment of the basin cavity—leading to tensional forces and the formation of the troughs.

5.6 Volcanism

Volcanism is repeatedly called upon to explain various features observed on Mercury and is inferred to have occurred throughout its history. Yet very few features, such as domes, cones and shield volcanoes, can be unequivocally identified as volcanic. Schultz (1977) and Malin (1978) have both considered volcanic features as might be seen on the Mariner 10 images and concluded that it would be very difficult to make positive identifications. For example, Malin examined images of volcanic features on the Moon obtained at resolutions comparable to Mariner 10 and found that their identification was far from definitive.

Malin did, however, suggest that some positive relief features could be volcanic domes (Fig. 5.30), although these features could be basin massifs. In addition, Dzuri-

Figure 5.29 Smooth plains of the Caloris basin showing troughs transecting ridges. Image width is about 50 km; north is toward bottom (Mariner 10 FDS 106).

Figure 5.28 Interior of the Caloris basin showing possible lava channel in the smooth plains (P. D. Spudis, personal communication, 1983). Image is about 30 by 100 km (Mariner 10 FDS 528996).

Figure 5.30 High-resolution image of Discovery Dome (arrow), a 15 km diameter feature which may be a volcanic dome (Malin 1978), or remnants of a basin rim.

sin has suggested that segments of some ridges may be volcanic features and inferred that the fractures associated with ridge and scarp formation could have served as avenues for magma extrusion.

However, until better images are available, the evidence for volcanism on Mercury remains inconclusive.

5.7 Geologic history

Analyses of the geologic and terrain mapping and consideration of the tectonic and thermal history exhibited by various landforms, such as arcuate scarps, lead to a synthesis of the evolution of the mercurian surface (from earliest to latest; modified from Dzurisin (1978)).

1. The formation of a solid lithosphere occurred prior to the end of heavy impact bombardment. The crust of Mercury may have been semi-molten and may not have recorded all impact events that were taking place; this may account for some of the intercrater plains and the possible lack of craters in some size ranges. Tidal spindown and/or crustal contraction (resulting from either lithospheric cooling or shrinkage of an iron-rich core) generated early-stage global tectonic fractures.

2. Continued crustal contraction resulted in compression to form arcuate scarps, some of which post-date large craters (as evidenced by cross-cutting relations), and the rejuvenation of early-formed tectonic patterns to generate lobate scarps. This period was accompanied by extrusion of volcanic materials to form additional intercrater plains. Some ridges and scarps may have served as avenues for magmatic extrusion.

3. The final stage of heavy impact cratering was marked by the formation of the Caloris basin. Ejecta from the basin produced radial grooves and ridges extending outward from the basin rim. Various smooth plains were emplaced both inside and outside the basin following the impact. Irregular ridges formed on the interior smooth plains as a result of compression, possibly as a consequence of basin floor subsidence, which occurred in response to magma withdrawal and extrusion of lavas to form the smooth plains outside the basin. Isostatic adjustment of the basin interior resulted in central uplift and the formation of extensional features to form troughs.

   Concurrent with the impact of the Caloris basin was the formation of the hilly and lineated terrain in the region antipodal to Caloris. This terrain is considered to have resulted from focusing of seismic energy resulting from the impact.

4. With the decline in the Solar System flux of bolides, Mercury experienced a gradually decreasing rate of impact cratering, similar to that observed on the Moon.

5. The youngest recognizable event in the evolution of the mercurian surface (exclusive of the continued formation of small impact craters) was the emplacement of smooth plains deposits. Ridges in some of the plains suggest an origin similar to lunar mare plains, that is, related to basaltic flood volcanism.

6 Venus

6.1 Introduction

Even to the casual observer, Venus is a distinctive planet. After the Sun and the Moon, it is the brightest object in the sky and has been called both the "evening star" because of its prominence after sunset, and the "morning star" because of its appearance during the dawn hours. However, even in ancient times, Venus was set apart from stars because of its "wandering," or planet-like, motion. Unlike most planetary objects, Venus rotates on its axis in a retrograde manner; that is, it spins in the direction opposite to its orbit around the Sun. The rate of rotation is so slow, however, that one day on Venus is equal to more than 116 Earth days.

In 1610, Galileo recognized, with the aid of his telescope, that Venus has lunar-like phases. This discovery lent support to the Sun-centered concept of the Solar System proposed by Copernicus. With the improvements in telescopes over the years, bright and dark markings were seen in the atmosphere of Venus, and by the mid-1880s observers noted that Venus has a gray halo when silhouetted against the Sun. Lomonosov, a Russian astronomer, used this information to infer the presence of an atmosphere on Venus. At about the same time, estimates of the diameter of Venus were made, which led to the idea of Venus and Earth being "twin" planets. This, in turn, aroused speculation that Venus was hospitable to life.

Visual and photographic spectroscopy were applied to Venus in the 1930s and resulted in the tentative identification of carbon dioxide as the principal component of the atmosphere. Kuiper (1962) extended these observations into the infra-red portion of the spectrum and found nearly 40 CO_2 absorption bands. By this time the presence of an extensive atmosphere, including clouds, was well established. Although Michael Belton and co-workers (1968), using infra-red spectroscopy, ruled out the possibility that the clouds were water vapor or ice crystals, microwave observations made in the 1960s and some early Soviet Venera results (Table 6.1) showed the presence of some water vapor, but not in sufficient quantities to account for the clouds. Results from microwave observations also yielded brightness temperatures of about 600 K at wavelengths greater than 3 cm, and it was inferred that these could result from either high surface temperatures or ionospheric effects. Measurements from Mariner

2 resolved this problem by showing that the surface is hot. At about this same time, Sagan (1961) and Pollack and Sagan (1965) formulated the **greenhouse model** for Venus in which solar energy penetrates the venusian atmosphere, is reflected from the surface and is re-radiated from clouds back to the surface, entrapping infra-red radiation. This could explain Venus's high surface temperature.

Until the mid-1970s, most studies of Venus concentrated on the evolution of the atmosphere and various astronomical considerations such as the period of rotation. With the landings of Venera spacecraft in the 1970s, interpretations of the composition of surface materials were made and images were taken of the surface. In addition, Earth-based radar observations shed light on global surface properties which, in combination with the Venera results, led to the initiation of geologic studies. Geologic and geophysical analyses were further enhanced with data obtained from the Pioneer Venus mission beginning in 1978. Although designed primarily to obtain atmospheric data, the Pioneer Venus orbiter carried a radar altimeter on board to measure radar "roughness" and to determine global topography, which allowed interpretations of the geophysical characteristics of Venus. In the mid-1980s, the Soviet Venera 15/16 orbits provided radar maps for part of the northern hemisphere of Venus. The early 1990s saw the successful operation of the United States' Magellan radar orbiter which mapped Venus globally and provided altimetric and other radar-derived data.

The combination of Earth-based radar, Pioneer Venus altimetry, results from the Venera missions, and, more importantly, information from the Magellan mission has generated speculation on the general geology of Venus and has placed constraints on various geophysical models of its interior. Because Venus is so like Earth in its size, density and position in the Solar System, this "twin" of Earth remains of high interest in comparative planetology. Hunten and co-editors (1983) provide an excellent collection of papers on all aspects of Venus known through the time of the Venera 13 and 14 landers in 1982; Barsukov *et al.* (1992) summarized results from the Soviet missions. Reports from the Magellan mission are given in the August and October 1992 issues of the *Journal of Geophysical Research (Planets)*.

Table 6.1 Missions to Venus.

Mission	Arrival	Type	Encounter characteristics
Mariner 2	14 Dec. 1962	Flyby	Closest approach: 34,833 km
Venera 4	18 Oct. 1967	Bus	Burn-up
		Entry probe	Hard lander, nightside
Mariner 5	19 Oct. 1967	Flyby	Closest approach: 4100 km
Venera 5	16 May 1969	Bus	Burn-up
		Entry probe	Hard lander, nightside
Venera 6	17 May 1969	Bus	Burn-up
		Entry probe	Hard lander, nightside
Venera 7	15 Dec. 1970	Bus	Burn-up
		Entry probe	"Soft" lander, nightside
Venera 8	22 July 1972	Bus	Burn-up
		Entry probe	"Soft" lander, dayside
Mariner 10	5 Feb. 1974	Flyby	Closest approach: 5700 km Images of cloud top
Venera 9	22 Oct. 1975	Orbiter	Periapsis: 1560 km; apoapsis: 112,200 km; period: 48 hours, 18 m; inclination: 34° 10'
		Entry probe	"Soft" lander, dayside
Venera 10	25 Oct. 1974	Orbiter	Periapsis: 1620 km; apoapsis: 113,900 km; period: 49 hours, 23 m; inclination: 29° 30'
		Entry probe	"Soft" lander, dayside
Pioneer Venus 1	4 Dec. 1978	Orbiter	Periapsis: <200 km; apoapsis: 66,000 km; period: 24 hours; inclination: 105°
Pioneer Venus 2	9 Dec. 1978	Bus	Burn-up, dayside
		Entry probes	4 hard landers, dayside and nightside
Venera 11	25 Dec. 1978	Flyby	Closest approach: 25,000 km
		Entry probe	"Soft" lander, dayside
Venera 12	21 Dec. 1978	Flyby	Closest approach: 25,000 km
		Entry probe	"Soft" lander, dayside
Venera 13	March 1982	Flyby	"Soft" lander with imaging Dayside
		Entry probe	
Venera 14	5 Mar. 1982	Flyby	"Soft" lander with imaging Dayside
		Entry probe	
Venera 15	10 Oct. 1983	Orbiter	Radar imaging
Venera 16	14 Oct. 1983	Orbiter	Radar imaging
Vega 1	June 1985	Balloon	Atmospheric data
Vega 2	June 1985	Balloon	Atmospheric data
Galileo		Flyby	Images and near-infra-red data on clouds
Magellan	August 1990	Orbiter	Radar imaging, radiometry altimetry, gravity

6.2 Radar data for Venus

Information on the geomorphology of Venus is derived primarily from radar observations. As discussed in Chapter 2, radar images differ markedly from optical images and require knowledge of radar geometry and remote sensing for proper interpretation. Radio waves (including radar) emitted or reflected from planetary surfaces yield a variety of information, including estimates of: (a) surface roughness (e.g. rocks, lava textures), shown by radar-bright images which typically indicate "rough" (or high radar backscattering) surfaces and radar-dark images which typically indicate smooth surfaces; (b) average surface slopes, indicated by root mean square (r.m.s.) values at scales ~10 times the radar wavelength; and (c) other physical and chemical characteristics such as the dielectric constant (which is a function of many complex, interrelated parameters, including surface composition).

Earth-based radar observations of Venus which began in the 1960s have steadily improved in quality. Most of these observations have been obtained from four facilities: (a) the Arecibo radar system in Puerto Rico which has an antenna 300 m in diameter and operates at wavelengths of 12.6 and 70 cm; (b) the deep-space tracking antenna at Goldstone, California, which has a diameter of 64 m and operates at a wavelength of 12.6 cm; (c) the Haystack radar system in Massachusetts which has an antenna diameter of 43 m and operates at 3.8 cm; and (d) the Russian Academy of Science which operates a 39 cm wavelength system. In general, the larger the antenna, the higher the spatial resolution. Thus, the Arecibo system can obtain data for Venus at about 1.3 km per radar pixel. Under optimum conditions, about 30 percent of Venus might be imaged by radar at a spatial resolution of ~2 km, but because of the orbital geometries of Earth and Venus, the area imaged is concentrated on either side of the equatorial region on the hemisphere that faces Earth during closest approach.

The Pioneer Venus spacecraft was placed into orbit in 1978 and data from the Pioneer Venus radar altimeter (Pettengill et al. 1980) have enabled the gross topography of Venus to be derived (Figs. 6.1, 6.2). These data cover the region between 74°N to 63°S and include about 93 percent of the surface with an elevation accuracy of about 200 m, integrated over an area about 100 by 100 km. In addition, coarse radar images were generated from the same radar system for the region between 45°N to 15°S and were used with other radar data to portray the general relief of Venus.

During the 1980s, Earth-based radar telescopes and the Soviet radar-mapping spacecraft, Veneras 15 and 16, returned images with an 8 cm wavelength system that provided clues to the geological diversity of Venus, as summarized by Basilevsky and Head (1988). But it was

Figure 6.1 Generalized topographic map of Venus showing elevations of surface features; contour interval 1 km (courtesy US Geological Survey).

Figure 6.2 Shaded relief map of Venus based on altimetric data from the Pioneer Venus orbiting spacecraft (courtesy US Geological Survey).

not until the successful operation of the United States' Magellan spacecraft, beginning in 1990, that the incredible complexity of Venus was revealed (Saunders *et al.* 1991).

As reviewed by Pettengill *et al.* (1991), the Magellan spacecraft obtained SAR (Synthetic Aperture Radar) images at 12.6 cm wavelength and spatial resolution of between 120 and 300 m. In addition, data on emissivity and ground elevations were obtained. By the end of its second year of operation, more than 97 percent of the planet had been mapped, some of it by radar-stereo imaging in which the same terrain is viewed by right-looking and left-looking SAR.

6.3 Physiography

Although relatively flat over most of its surface, Venus has substantial topographic relief in some areas. Two features, named Alpha and Beta, were identified on the basis of their high radar backscatter properties in Earth-based radar images. One smaller radar feature, named Eve, is southwest of Alpha and is a radar-bright ring encircling an intense radar-bright central spot. This feature is used as the planet-fixed reference for longitude; because Venus is in retrograde rotation, longitude increases toward the east by International Astronomical Union convention. Except for Alpha, Beta and a few other features, most geographic features on Venus are given women's names such as those of goddesses from mythology.

The highest point on Venus occurs within Maxwell Montes (the only feature named after a man) and rises at least 11.5 km above the datum (datum is based on the mean planetary radius of 6051.9 km). For comparison, Mt. Everest rises about 12 km above the Earth's median elevation (which is about 3 km below sea level). The lowest area on Venus is in Diana Chasma, a trench centered at 14°S, 156°E. The floor of the trench is at least 2 km below datum, equivalent to the Dead Sea Rift on Earth; however, it is less than one-fifth the depth of the Mariana Trench which at −8 km below the Earth's median elevation (−11 km below sea level) is the lowest place on Earth. Thus, the total topographic relief on Venus is about 13 km, considerably less than Earth's maximum relief of 20 km.

Hypsometric curves show the percentage of global topography lying at different elevations and are useful for planetary comparisons. As shown in Figure 6.3, Venus is strongly unimodal in its distribution of topography, Earth is strongly bimodal, and Mars is trimodal. The bimodal distribution of topography on Earth results from two distinct zones; the continental platforms (including the continental shelf) and the deep-sea basins. Early workers, including G. K. Gilbert and Alfred Wegener, speculated that if Earth's crust consisted of homogeneously distributed materials, then the hypsographic curve should be a simple Gaussian distribution. However, recognition of two distinct topographic levels led them to consider the compositional and density differences between continental crustal materials (sial) and oceanic crustal materials (sima) and the consequences of isostatic adjustments. Except for the maximum relief noted above, Venus is

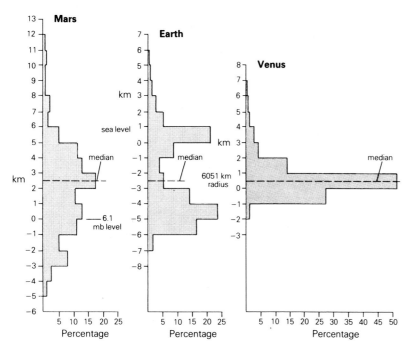

Figure 6.3 Hypsographic curves for the terrestrial planets, showing Venus (unimodal distribution), Earth (bimodal distribution reflecting continent and ocean basins), and Mars (trimodal distribution for the northern lowlands, the intermediate-level cratered terrain, and the high volcanoes).

extremely flat, with 60 percent of the surface falling within 500 m of the modal planetary radius, leading some investigators to speculate that the venusian crust may not be differentiated in a manner similar to Earth's crust.

Masursky and colleagues (1980), using Pioneer Venus data, subdivided Venus into three provinces based on elevation: (a) *upland rolling plains* which lie between 0 and +2 km elevation and constitute 65 percent of the surface; (b) *highland provinces* which are higher than +2 km elevation and constitute 8 percent of the surface; and (c) *lowland areas* which are lower than the 0 km datum and constitute 27 percent of the surface. While these provinces are based solely on elevation, there is also a general correlation between elevation and radar-brightness (i.e. surface roughness): the upland rolling plains are of intermediate brightness, the highland provinces are radar-bright, and the lowlands are radar-dark. This correlation may be related to surface processes in which high-standing areas are eroded and fine-grained materials are removed, leaving blocky surfaces. These fine-grained sediments are then deposited in the lowlands to produce smooth, radar-dark surfaces.

Owing to the topographic relief of the surface, atmospheric temperature and pressure vary over the various topographic provinces. In the lowlands, temperatures are approximately 465°C and atmospheric pressure is 96 bar; at the elevation of the highlands, the temperature and pressure are lower: 374°C and 41 bar, respectively (Warner 1983). This wide range of conditions means that impact, volcanic, weathering, and mass transport processes may operate differently in the highlands and the lowlands.

6.4 Craters

Impact craters have been identified on all the other terrestrial planets and their presence was expected on Venus. Early Earth-based radar images revealed numerous circular features thought to be impact craters (Cutts *et al.* 1981). Radar images obtained by the Venera 15/16 spacecraft early in the mission revealed distinctive impact craters with ejecta and numerous features that were interpreted to be multi-ringed impact basins. Magellan results (Phillips *et al.* 1991) show a wide diversity of impact scars and related features, including some which appear to be unique to Venus. However, the number of impact craters found on Venus is low when compared to Mercury and the Moon, suggesting that much of the venusian surface is geologically young, perhaps no more than 500 million years (Schaber *et al.* 1992). Moreover, most of the craters appear to be about the same age—an observation that is difficult to explain, at least in comparison to

what is known about the histories of other terrestrial planets and satellites.

As is typical for impacts on terrestrial planets, large events result in multi-ringed craters, as seen in the 150 km diameter crater Meitner (Figs. 6.4, 6.5) whereas smaller craters show central peaks or peak clusters and complex rim structures. With decreasing size, these craters give way to relatively simple bowl-shaped craters.

The high temperature and dense atmosphere on Venus (Table 2.1) may influence impact cratering processes in several ways: (a) by slowing the incoming bolide; (b) by influencing the emplacement of some ejecta (Fig. 6.6); and (c) by enhancing post-impact modifications of the crater. Tauber and Kirk (1976) assessed the effects of the venusian atmosphere on impacting bolides and noted that small objects will break apart before impacting the surface. Fracturing in-flight would be caused by frictional heating and/or air pressure acting directly on bolides at high speeds. They calculated that the smallest stony meteoroid that would survive is ~ 80 m in diameter, whereas the smallest iron meteoroid would be 30 m. From scaling relationships based on impact kinetic energies, Tauber and Kirk estimated that the smallest craters would be 300–400 m for stony bolides and 150–200 m for iron bolides, thus placing lower limits on the sizes of impact craters expected on Venus.

Magellan images indicate the influence of the atmosphere on small oncoming bolides; clusters of small craters are thought to represent bolides that have been dis-

Figure 6.4 Magellan radar image of Meitner crater, an impact structure 150 km in diameter located at 55.6°S, 321.6°E, showing multiple rim and ejecta field (from Schaber *et al.* (1992)).

Figure 6.5 Dickinson crater, an impact structure 69 km in diameter at 74.6°N, 177.3°E showing central peak cluster, complex rim and ejecta, and radar-dark floor (JPL P-39716).

Figure 6.6 Radar mosaic showing complex ejecta deposits associated with impact crater Carson, centered at 25°S, 345°E; parabolic dark zone is suggested to be fine-grained ejecta whose distribution was influenced by westward-moving (toward the left) winds at the time of emplacement. Dark rectangular areas are gaps in data acquisition. Area shown is about 410 km across (Magellan JPL MRPS 42369).

rupted by the atmosphere to produce a shotgun-like pattern of impact structures. Other small objects appear to be so disrupted that they leave only a radar-dark scar on the surface, with no apparent craterform (Figs. 6.7, 6.8).

Ejecta deposits are visible around most venusian craters. However, the atmosphere also affects the emplacement of impact ejecta. Based on calculations of atmospheric drag, Tauber and co-workers (1978) noted that the range of some ejecta fragments would be substantially less on Venus than on planets with thin or no atmospheres. Their calculations, however, did not take into account the possibility of turbulent flow within the ejecta cloud, and Schultz and Gault (1979) and Schultz (1981, 1992) have speculated that ejecta fragments <10 m could be carried along the surface as entrained flow. They also suggest that ejecta from large (>100 km) craters might not be slowed appreciably upon ejection, but during re-entry fragments as large as 10 m would be decelerated so that fragmentation upon impact would be minimal.

The high temperature on Venus is expected to influence the amount of impact-generated melt and vaporization, as discussed by Grieve and Head (1982). They estimate that as much as 50 percent more melt and vapor would be produced on Venus than a comparable impact on Earth. These predictions were confirmed by Magellan images which reveal flow-like ejecta associated with some impact

Figure 6.7 Radar-dark feature in the Sedna Planitia region (40°N, 15°E) thought to be an impact scar resulting from the extensive break-up by the atmosphere of in-coming bolide; area shown is about 290 km across (Magellan C1 MIDR 45N350; from Schaber *et al.* (1992)).

140

Figure 6.8 Radar-bright feature in Sedna Planitia representing impact scar from disrupted bolide; area shown is about 275 km across (Magellan C1 MIDR 30N009; from Schaber *et al.* (1992)).

Figure 6.10 Radar picture showing channellike feature 150 km long which appears to be associated with ejecta from Aglaonice impact crater in the Lavinia region of Venus near 27°S, 339°E. Area shown is 200 km × 250 km (part of JPL P-36711).

craters. As shown in Figure 6.9, in some cases the ejecta deposits extend several crater diameters from the crater rim and have the appearance of a fluidized mass. These masses have very high radar backscatter cross-sections, suggesting that they have extremely rugged surfaces. Moreover, channels as long as 150 km are found in association with the ejecta deposits (Fig. 6.10). The channels have sinuous walls 100 or more meters high, show island-like features on their floors, and empty imperceptibly onto the surrounding plains. As discussed by Schaber *et*

al. (1992), whether these channels were formed by melted rock generated as part of the impact process, or are a form of volcanism "triggered" by the impact, or resulting from some totally unknown process is a matter of debate that is yet to be resolved.

Figure 6.9 Impact crater Addams (90 km in diameter; 56.1°S, 98.9°E) showing bright outflow ejecta that extends 600 km from crater rim; ejecta is thought to include impact melt. (Magellan F-MIDR 55S097; from Schaber *et al.* (1992)).

Figure 6.11 Magellan image of a corona structure some 250 km across located in plains south of Aphrodite Terra at 59°S, 164°E. Several flat-topped domes, some of which are superposed on the corona ring fractures, suggest that magma was associated with the formation of the corona (JPL P-38340).

Figure 6.12 Radar image of the Lakshmi region, centered at 30°N, 333°E, showing deformation of the crust by two sets of fractures. The prominent set trends northwest–southeast, with less prominent fractures intersecting at about right angles, breaking the terrain into rectangular blocks about one by two kilometers in size. Area shown is about 37 km × 80 km (JPL P-36699).

142

6.5 Tectonic features

Because Venus is nearly the same size and density as Earth, and because both planets are close neighbors in the Solar System, the two are often considered "sister" planets. Theories of the interior configuration of Venus based on its global properties, the lack of a magnetic field, and geophysical measurements made from spacecraft all suggest that Venus has a lithosphere less than 100 km thick overlying an upper mantle, and a core that may be about the same size as Earth's core. In terms of geological processes, such as volcanism and tectonism, it is reasonable to suppose that Earth and Venus might share some common attributes. Consequently, one of the key goals for the exploration of Venus was to search for evidence of Earth-like plate tectonics. For example, does Venus show evidence of crust that has spread apart, lithospheric plates that have sheared, and hot spots? Magellan results show no indications of these features, other than hot spots and a local area of possible subduction. The best evidence for possible hot spots is in the form of huge circular features seen on the surface termed **coronae,** some of which are more than 600 km across. As shown in Figure 6.11, coronae consist of discontinuous ridges and grooves arranged in concentric patterns that might represent deformation of the crust by upwelling from the venusian mantle (Squyres *et al.* 1992; Stofan *et al.* 1992). However, models of the interior suggest that the upper mantle of Venus may be rather stiff, inhibiting convection. Thus, an alternative explanation is that coronae may be related to impact processes and that they represent highly degraded or deformed craters formed early in Venus' history.

As noted by Sandwell and Schubert (1992), some coronae, such as Artemis and Latona, have annular moats that are similar to the topography of trenches and outer rises found in association with subduction zones on Earth. Based on geophysical modelling, they proposed that lithospheric subduction may be occurring on the margins of some large venusian coronae.

Venus seems to have its own unique set of tectonic features. For example, some areas are referred to as *tesserae* terrain (from the Greek word, *tessares,* referring to four corners). Tesserae terrain consists of crust fractured into blocks a few kilometers to 20 km across and is evidence for extensive tectonic deformation (Fig. 6.12). In other places, belts of ridges and grooves hundreds of kilometers long (Fig. 6.13) may reflect large-scale tectonic deformation not apparently related to Earth-like plate motion.

6.6 Volcanic features

Volcanic features are abundant on Venus (Head *et al.* 1992; Guest *et al.* 1992). Some of the large mountains are shield volcanoes (Figs. 6.14, 6.15) as much as several hundred kilometers across which stand 4 km high and are composed of accumulations of lava. Some volcanoes are associated with an enormous rift valley that can be traced more than 3000 km, comparable to the rift valley of east Africa (Fig. 6.16). Other areas, such as Ishtar

Figure 6.13 Magellan radar image of the northern boundary of Ovda Regio, centered at 1°N, 81°E showing linear ridge terrain. The dominant features in the image are low-relief, rounded, linear ridges, typically 8–15 km wide and 30–60 km long. Radar-dark material, either volcanic or aeolian in origin, fills the regions between some ridges; area shown is 380 km across (Magellan F-MIDR 00N082; from Solomon *et al.* (1992)).

Figure 6.14 Large shield volcano 480 by 355 km in diameter located west of Phoebe Regio, centered at 3.5°, 270° (Magellan C1 MIDR 00N266; from Head *et al.* (1992)).

Figure 6.15 Shield volcano 40 by 60 km, centered at −7.5°S and 200.5°, showing numerous lava flows originating from a central vent and fissure (Magellan F-MIDR 10S200; from Head *et al.* (1992)).

Terra, are punctuated by huge calderas. In addition, fields of small volcanic domes, cones, and shields are found in many places on Venus and attest to local eruptions.

Many of the vast expanses of plains on Venus are revealed by radar images returned by the Magellan spacecraft to be covered by sheets of lava (Fig. 6.17), some of which were erupted from fissures. Some of these flows were emplaced by rivers of lava flowing through channels hundreds or even thousands of kilometers long (Fig. 6.18). It is difficult to imagine lavas remaining sufficiently fluid over such long distances, even in the hot venusian environment. Some investigators speculate that somewhat exotic (by Earth standards) lava compositions, such as komatite or carbonatite, might be involved (Baker *et al.* 1992).

As a consequence of magma chambers spilling their lavas onto the surface through volcanic eruptions, some parts of the crust collapsed, leaving irregular depressions and pit craters on the surface. One such area is seen north of Ovda Regio. Shown in Figure 6.19, this area displays linear depressions and collapse pits, some of which apparently "fed" lava into sinuous channels.

Viscous lava flows and domes are also present on Venus (Fig. 6.20). In addition, Venus has a type of volcano not seen elsewhere in the Solar System. These volcanoes, some of which are shown in Figure 6.21, consist of circular domes or flat-topped mounds of lava presumed to have been rather viscous and pasty at the

time of their eruption (Pavri *et al.* 1992). They are as large as 25 km across and hundreds of meters high, and are thought to represent a style of eruption or a type of magma not previously encountered in planetary exploration. The flanks of some of these volcanoes appear to have been modified, perhaps by landslides.

Does Venus currently experience active volcanism? Observations of large variations of sulfur dioxide (a common product of active volcanoes on Earth) in the atmosphere lead some planetary scientists to think so. The relatively fresh lava flows visible on Magellan images could also support the notion of geologically-recent volcanism. But, as yet, no direct evidence of active volcanoes has been found.

The Soviet Venera lander spacecraft provided clues to the composition of the lava flows and other rocks on Venus (Figs. 6.22, 6.23). Most of the landings occurred on the flanks of Beta Regio, a prominent volcanic region astride the equator. Measurements made by instruments on the landers show that the rocks are similar to basalts found on the sea floor of Earth. However, a rare (for Earth) kind of basalt with a high percentage of potassium, and a basalt rich in sulfur were also identified. Two other Soviet probes, each from the Vega mission, landed in Rusalka Planitia on the northern flanks of Aphrodite Terra and also returned measurements typical for basaltic rocks. Thus, as has been found on Earth, Earth's Moon, and Mars, eruptions of basalt and basalt-like lava flows seem to have been common on Venus.

Figure 6.16 Sketch map of (a) the Beta Regio–Phoebe Regio area, based on radar images, compared to (b) the East Africa Rift; similarities in size, form and geometric relationships led McGill *et al.* (1981) to consider this region of Venus to result from mantle upwelling, tectonic deformation of the crust and eruptions to produce superposed volcanoes (from McGill *et al.* (1981) *Geophys. Res. Let.,* copyright American Geophysical Union).

Figure 6.17 Magellan radar mosaic centered at 55°S, 354°E showing an extensive lava flow field, known as Mylitta Fluctus, in Lavinia Planitia. The entire flow field covers an area 800 km × 380 km. The dark band is an area not imaged on one of the spacecraft orbits (JPL P-38292; from Roberts *et al.* (1992)).

145

Figure 6.18 Sinuous lava channel abotu 2 km wide found on lava plains at 49°S, 273°E; area shown is 130 by 190 km (JPL P-39226).

6.7 Surface modifications

In addition to impact cratering and deformation by volcanism and tectonism, the surface of Venus has been altered by erosion. In the searing venusian environment, liquid water cannot exist on the surface. If water has not modified the surface of Venus, is there evidence of other geological processes of erosion? Venera lander photographs (Fig. 6.24) and Magellan radar images obtained from orbit suggest that some weathering and erosion does take place. The dense, acid-rich, high temperature atmosphere may be corrosive. Sediments and other fine-grained material, at least partly weathered from rocks, are seen in all four of the landing sites where the Venera spacecraft obtained pictures. The amount of sediments, however, differs from one site to another.

Although only sluggish winds of 0.5 to 2 meters per second have been measured on the surface of Venus, this is sufficient to move sand grains in the high-density atmosphere (Fig. 3.36; Greeley and Arvidson, 1991). Radar data suggest that most of the venusian surface is bedrock, with only about one-fourth of the planet being covered with porous or loose materials. Consequently,

there may be insufficient agents of abrasion, such as sand grains, to cause significant erosion of the surface (Arvidson *et al.* 1992). Nonetheless, Magellan images do show evidence of the work of wind in some regions (Greeley *et al.* 1992). For example, Figures 6.25 and 6.26 show fan-shaped bright (radar-reflective) features called *wind streaks* up to 20 km long that are associated with small hills. Most of the surface around the hills is only moderately radar reflective, possibly indicating the presence of loose particles. The bright zones are considered to be places where winds streaming around the hills have stripped the loose material from the surface. This has exposed bedrock that is radar reflective and shows as a bright zone on the radar pictures. Wind streaks are found in several places on Venus and serve as local "wind vanes", recording the prevailing wind direction at the time of their formation. More than 8000 wind streaks have been found on Venus. Mapping their orientation reveals near-surface winds probably influenced by a Hadley cell circulation.

In addition to wind streaks, several areas on Venus show features that look like sand dunes on Earth. These features are a few hundred meters wide and occur in fields

146

Figure 6.19 Venusian channels located just north of Ovda Regio. The radar image is centered at 11.4°S, 89.5°E and is approximately 110 km wide. Note the enlarged fractures and elongated collapse regions for source areas and the sharply defined channel walls (Magellan F-MIDR 10S087; from Baker *et al.* (1992)).

Figure 6.20 Magellan image of a lava flow field on plains between Artemis Chasma and Imdr Regio at 37.5°S, 164.5°E. This volcanic structure is composed of numerous flows, some more than 500 m thick; note the "festoons" on many of the flows; area shown is 300 km across (JPL P-39916).

Figure 6.21 Volcanic domes as large as 65 km across and several hundred meters thick in the Eistla region at 12.3°N, 8.3°E. These and similar domes on Venus are considered to have been formed by lavas of high viscosity (JPL P-38388; from Moore *et al.* (1992)).

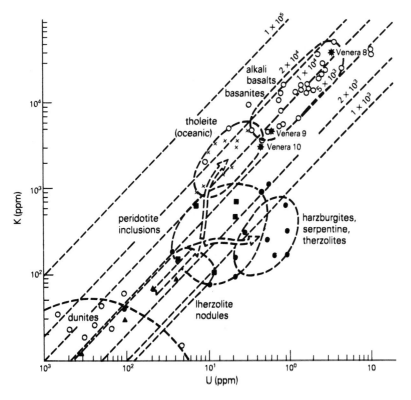

Figure 6.22 Diagram for the content of potassium and uranium in rocks on Earth compared with results obtained by Veneras 8, 9, and 10 (from Surkov (1983)).

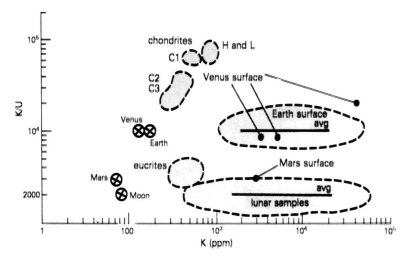

Figure 6.23 Diagram of potassium abundances and K/U ratios for Venus, Earth, Mars, Moon, and selected meteorites showing the similarity between Venus and Earth (from Taylor (1982)).

148

a ВЕНЕРА-9 ОБРАБОТАННОЕ ИЗОБРАЖЕНИЕ

b ВЕНЕРА-10 ОБРАБОТАННОЕ ИЗОБРАЖЕНИЕ

c ВЕНЕРА-13 ОБРАБОТКА ИППИ АН СССР И ЦДКС

d ВЕНЕРА-14 ОБРАБОТКА ИППИ АН СССР И ЦДКС

Figure 6.24 Venera lander images: (a) Venera 9, (b) Venera 10, (c) Venera 13, and (d) Venera 14 (images courtesy of Jet Propulsion Laboratory).

Figure 6.25 Magellan radar image of the area northeast of Ushas Mons, Venus, showing bright zones, called *wind streaks*, formed downwind from small volcanic cones. These wind streaks are about 10 km long and are thought to indicate the prevailing wind direction (from the top to the bottom of the picture) at the time of their formation; area shown is 40 km by 112 km (JPL P-36698).

Figure 6.26 Magellan images of a radar-bright wind streak 25 km long associated with a small (5 km) volcano at the western end of Parga Chasm at 9.4°S, 247.5°E. The wind is inferred to have been blowing from the left toward the right at the time of streak formation (JPL P-38810).

Figure 6.27 Sand dunes and radar-bright wind streaks in the Fortuna-Meshkenet dune field, centered at 67.7°N, 90.5°E (from Greeley *et al.* (1992)).

nearly one hundred kilometers across (Fig. 6.27). One such field is found in association with ejecta deposits of a large impact crater. There is speculation that dune fields can form on Venus wherever sufficiently strong winds and a supply of sand grains occur, and both conditions are apparently met around some craters where impact ejecta deposits provide a ready source of sand.

Landslides of several forms occur on Venus (Malin, 1992), and represent another mechanism for modifying the surface. Landslides are found on slopes in some tessera terrain where relatively steep slopes occur, enhancing the conditions for mass wasting; other landslides occur on the flanks of some dome volcanoes.

6.8 Geologic history

Geologic histories for planets and satellites are derived primarily from the ages and distributions of rocks seen on their surfaces, coupled with knowledge or inferences about their interior characteristics. Unfortunately, information for Venus is rather incomplete, particularly in comparison to most of the other terrestrial planets. Only the northern part of the planet has been mapped geologically. No seismic data (the primary source for information about planetary interiors) have been obtained. Moreover, because impact craters are sparse, regional age relationships are difficult—if not presently impossible—to determine. Consequently, knowledge of the geologic history of Venus is incomplete.

Like the other terrestrial planets, Venus was probably assembled from planetesimals and other small bodies

about 4.6 billion years ago. Sufficient heat was generated by collisions to completely melt the proto-planet. With time and cooling, a crust, mantle, and core probably developed from the melt. Because Venus and Earth are comparable in size and density (and, hence, probably similar in composition), the configuration of the interiors of both planets are thought to be similar.

The lack of abundant impact craters on Venus suggests that most of the crust that formed early in its history has been destroyed, either by burial by lava flows, tectonic "recycling" of the crust, or by erosion and burial of the surface by sediments. Of these three possible explanations, burial by lava flows seems to be the most likely because volcanoes, volcanic craters, and lava flows are abundant on Venus. Destruction of the impact craters by crustal plate "recycling" remains a possibility, but evidence of such recycling has not been found, at least not in a style similar to that of Earth.

The processes which dominate most of Venus' history are volcanism and tectonism. The formation of both large and small volcanoes and the emplacement of vast sheets of lava occurred over long periods of time on Venus, and were accompanied by deformation of the crust. Both of these processes are manifestations of the loss of heat from the interior of the planet. It is possible—perhaps even likely—that these processes are active today. Are other processes active? Sluggish winds do sweep the surface with sufficient energy to move loose sand and dust, but the amounts of material available for transport by the wind may be rather limited. Impact cratering can also occur today just as it does on other planets. But because of the dense atmosphere on Venus, only the larger objects will reach the surface.

Perhaps the most puzzling aspect of Venus' history is the evolution of its atmosphere. One might have expected the atmosphere of Earth and Venus to be similar, given the other similarities of these "sister" planets, and perhaps they were early in their histories. Yet the evolution of the atmospheres on the two planets took drastically different paths at some point in their geologic evolution, leading to Earth with its mild, nitrogen-oxygen atmosphere, and Venus with its dense, carbon dioxide atmosphere. Many ideas have been offered to explain the reasons for these differences. For example, like Venus, Earth also evolved a large quantity of carbon dioxide. Today most of the carbon dioxide on Earth is contained in limestone and other rocky deposits, formed primarily by precipitation from sea water. If all of the carbon dioxide presently contained in these rocks were released, Earth, too, would be enveloped in a thick carbon dioxide atmosphere. And it is likely that surface temperatures would rise through the greenhouse effect, just as they are thought to have risen on Venus.

Thus, part of the difference between Earth and Venus is linked to the development of oceans on Earth. Based on geochemical evidence, some planetary scientists have suggested that Venus, too, may have evolved sufficient water to have had oceans, oceans long since boiled away if they ever did exist. One of the puzzles of Venus is to determine if large quantities of water were present early in its history. The geological exploration of Venus is still in its infancy. Until analyses of Magellan data are refined and a more global, uniform view is obtained, knowledge of the geological history for Venus remains fragmentary.

7 Mars

7.1 Introduction

It is no wonder that the planet Mars is named for the Roman God of War. Its bright, blood-red color is in stark contrast to the night sky and must have inspired visions of battle to ancient observers. In the last hundred years or so, the Red Planet, like Venus, has evoked speculations on bizarre lifeforms, advanced civilizations, and Earth-like climates ranging from cold and wet to hot and dry.

These speculations were based on observations combined with fertile imaginations. When seen through telescopes, Mars displays distinctive surface patterns (Fig. 1.1(a)), some of which change with martian seasons. For example, white patches in the polar regions shrink toward the poles in the summer and a dark zone, called the "wave of darkening", extends toward the equator. This sequence was interpreted to represent the melting of ice caps and the subsequent release of water. In all, this is not an unreasonable interpretation based on pre-space age knowledge—the white patches are, in fact, polar caps, but they are composed mostly of CO_2 frost which ablates during the summer. The wave of darkening, however, may simply be a photometric effect of differences in illumination and viewing angles from season to season. Alternatively, it may be the result of contrast differences in which some areas become brighter, perhaps as a result of atmospheric dust.

Many imaginative science fiction themes for Mars arose from misuse of the term **canale**—the Italian word for "*channel.*" Apparently introduced in the 1860s by the Italian observer, Secchi, for some linear markings that he saw on Mars, Schiaparelli later used the term during his observations in 1877 and recorded them on his well-known map of 1878. "*Canale*" was transliterated by English-speaking workers to "*canal*" and touched off the notions of intelligent life. Enamored by this possibility, Percival Lowell, of a wealthy Boston family, built an observatory at Flagstaff, Arizona, for the primary purpose of observing Mars. He drew maps showing extensive canals and, around the turn of the century, Lowell published three books in which he proposed the existence of advanced civilizations on Mars and described the construction of elaborate irrigation network canals for the distribution of water over the whole planet. These and earlier reports inspired science fiction writers, including Edgar Rice Burroughs and H. G. Wells, who enter-

tained—and, in some cases, terrified—many a reader. The ideas of intelligently constructed canals and notions of advanced lifeforms were not put to rest until well into the space age.

Despite several successful early missions to Mars (Table 7.1), it was not until Mariner 9 that global knowledge of the surface, including the absence of canals, was obtained. Nonetheless, it was the quest for extraterrestrial life that stimulated the subsequent Viking missions. In-

Table 7.1 Successful missions to Mars.

Spacecraft	Encounter date	Mission	Event
Mariner 4	14 July 1965	Flyby	Closest approach 9912 km; 22 images.
Mariner 6	31 July 1969	Flyby	Closest approach 3330 km; 74 images.
Mariner 7	5 Aug. 1969	Flyby	Closest approach 3518 km; 125 images.
Mariner 9	13 Nov. 1971	Orbiter	6900 images; ultraviolet and infra-red spectrometers; infra-red radiometer.
Mars 3[*]	2 Dec. 1971	Orbiter Lander	20 sec of data returned.
Mars 5[*]	2 Feb. 1974	Orbiter	Transmitted images and served as relay for lander.
Mars 6[*]	12 Mar. 1974	Lander	Soft landed at 24°S, 25°W; returned atmospheric data during descent.
Viking 1	20 July 1976	Lander and orbiter	Lander obtained images, measured wind speeds, temperatures and directions; measured chemical and physical properties of the surface.
Viking 2	3 Sept. 1976	Lander and orbiter	Orbiters returned 55,000 images showing surface details as small as 10 m, collected gravity field data, monitored atmospheric water levels, thermally mapped selected surface sites.
Phobos 2[*]	Jan. 1989	Lander orbiter	Lost contact after initial return of orbiter data.

*Soviet missions.

Figure 7.1 (opposite) High-resolution image of the martian moon, Deimos, taken by Viking Orbiter 2 from a distance of only 50 km. Rocks as small as 3 m across are visible and are thought to be impact ejecta; illumination from the left JPL P-19599).

Figure 7.2 (below left) View of the martian moon, Phobos, showing grooved and cratered surface (Viking Orbiter 39B84).

volving two orbiters and two landers, all operating simultaneously, the Viking project remains the most complex unmanned mission in Solar System exploration. The primary objective of the project was to search for life on Mars. Although results from the biological experiments were essentially negative or at best ambiguous (Klein (1979) provides an excellent review), the Viking mission provided a wealth of data on the geologic, geophysical and atmospheric characteristics of Mars.

Mars has turned out to be a planet of superlatives: it has the biggest volcanoes, the largest impact basin, the biggest canyons and the longest channels of any planet or satellite that we have seen in the Solar System. Thus, while the science fiction enthusiasts may have been disappointed in the biological results, the geologic aspects have been exceedingly rewarding.

7.2 Phobos and Deimos

Mars has two small moons, Phobos (21 km in diameter) and Deimos (13 km in diameter). Both are heavily cratered objects, having crater frequency distributions similar to the Moon (Veverka and Thomas 1979). Viking Orbiter 2 flew as close as 28 km to Phobos and returned the highest resolution ($\sim$ 1m) pictures from the mission. Similar high-resolution images of Deimos show tiny craters and blocks of presumed ejecta (Fig. 7.1).

Images of Phobos (Fig. 7.2) show a curious pattern of grooves. As mapped by Thomas *et al.* (1978), these grooves radiate from a 10 km crater, Stickney, and converge on the opposite side of the satellite at the antipode to the crater, suggesting that the grooves are somehow related to the impact event.

The albedo, spectral reflectance and low density (<2.0 g cm^{-3}) of both moons suggest that they have a composition similar to carbonaceous chondrites (meteorites rich in water and organics). Because carbonaceous chondrites are considered to have formed in the asteroid belt, Pollack *et al.* (1979) suggested that Phobos and Deimos are captured asteroids.

7.3 Physiography

Mars exhibits a wide range of terrains (Fig. 7.3) and has considerable topographic relief. Place names are derived from classical mythology and from a system adapted by the International Astronomical Union (Blunck, 1977). In general, large martian craters are named for famous people, small craters (5–100 km) are named for small towns on Earth, sinuous valleys are named after "Mars" in various languages, and other landforms are named from regional albedo features in combination with a generic term, such as **mons** (mountain).

As shown by its hypsographic curve (Fig. 6.3), Mars has a three-fold topographic distribution: a mid-level terrain which has the greatest areal extent, the lowland plains and the mountains. Elevations are referenced to a 0-elevation plane which is based on atmospheric pressure. The pressure selected for datum is 6.1 mbar, which is the triple point for water (at lower pressures, liquid water is unstable) and is near the present mean pressure. The summit of Olympus Mons at 27 km marks the highest elevation on Mars while the floor of the Hellas basin is about −4 km elevation. Thus, the total relief on Mars exceeds 30 km, over a third greater than the total relief on Earth.

There is a general asymmetry in the distribution of the major terrains on Mars. The sparsely cratered lowland plains of the northern hemisphere are separated from the southern cratered highlands by a great circle inclined about 30° to the equator in the western hemisphere. Superimposed on the young, lowland plains are the elevated Tharsis and Elysium regions and their volcanoes. The reason for the lowland–highland dichotomy remains one of the major unsolved problems on Mars.

Data from the Mariner and Viking missions enabled the general physiographic provinces and terrain types on Mars to be defined, as shown in Figure 7.4 (terrain units are indicated by abbreviations in the text and on the map). The general geology of Mars prior to the Viking mission is given by Mutch and colleagues (1976) and summarized on a geologic map by Scott and Carr (1978). Excellent post-Viking discussions of the surface of Mars are provided by Arvidson et al. (1980), Carr (1980; 1981) and Cattermole (1992) and outlined on maps by the US Geological Survey. The book, *Mars,* published by the University of Arizona Press, 1992, summarizes the state of knowledge prior to the arrival of the Mars Observer spacecraft.

7.3.1 Basins

Circular basins of presumed impact origin constitute the oldest recognizable features on Mars. In addition to the Hellas, Argyre and Isidis basins, Wood and Head (1976) and Schultz and co-workers (1982) recognize numerous additional basins (Table 7.2; Fig. 7.4). Centered at 295°W, 40°S, the Hellas basin measures 1600 by 2000 km and has a rim that is 50–400 km wide, making it the largest impact basin in the Solar System. Concentric patterns of weakness extend as far as 1600 km from the center of Hellas. The floor of the Hellas basin appears to

Table 7.2 Impact basins on Mars; numbers are keyed to Figure 7.4 (from Wood and Head (1976) and Schultz et al. (1982)).

Designation	Location	Ring(s) diameter (km)
1. (overlapped by Newcomb Crater)	3°W, 23°S	380, 800
2. Aram Chaos	21.5°W, 3°N	140, 250, 440, 550
3. Ladon basin	29°W, 18°S	270, 470/580, 975
4. Holden basin	32°W, 25°S	260, 580
5. Chryse basin	47°W, 19°N	800, 1400, 2000, 2750, 3600, 4300
6. Mangala basin	147°W, 0°N	300, 570
7. Sirenum basin	166.5°W, 44°S	500, 1000
8. (arc of massifs in Arcadia)	167°W, 37°N	380, 600
9. (south of Hephaestus Fossae)	180°W, 30°S	180, 340, 1000
10. Al Qahira basin	190°W, 20°S	300, 780, 1200
11. (south of Hephaestus Fossae)	233°W, 10°N	500, 1000
12. (south-east of Hellas)	273°W, 58°S	225, 500
13. Nilosyrtis Mensae	282.5°W, 33°N	380
14. (south of Renaudot)	297°W, 38°N	600
15. (south of Lyot)	322°W, 42°N	60, 145, 260, 400, 480, 570
16. Cassini	328°W, 24°N	220, 450, 850
17. Deuteronilus B	338°W, 43°N	55, 201
18. Deuteronilus A	342°W, 44°N	44, 80, 220, 280
19. (overlapped by South Crater)	213°W, 73°S	300, 680
20. (overlapped by Schiaparelli)	346.5°W, 5°S	140, 560
21. (west of Le Verrier)	356°W, 37°S	430
22. Liu Hsin	171.5°W, 55°S	60, 135
23. Gale	222.0°W, 5°S	85?, 150
24. near Columbus	164.5°W, 25°S	70, 145
25. Ptolemaeus	157.5°W, 46°S	65?, 150
26. Phillips	44.5°W, 67°S	90?, 175
27. Molesworth	210.5°W, 28°S	80?, 180
28. Lowell	81.3°W, 52°S	90, 190
29. Lyot	330.5°W, 50°N	100, 200
30. Kaiser	340.5°W, 46°S	100?, 200
31. Kepler	218.5°W, 47°S	110, 210
32. Galle	31.0°W, 51°S	100, 220
33. Herschel	230.2°W, 15°S	160?, 290
34. Antoniadi	299.1°W, 22°N	200?, 400
35. Huygens	304.2°W, 14°S	250, 460
36. Schiaparelli	343.6°W, 3°S	230, 470
37. South Polar	267.0°W, 83°S	?, 850
38. Argyre	42.0°W, 50°S	560?, 1200
39. Isidis	272.0°W, 16°N	1100, 1900
40. Hellas	291.0°W, 43°S	?, 2000

South polar region

North polar region

Figure 7.3 Shaded airbrush relief map of Mars showing prominent features, selected place names, and Viking Landers 1 and 2 sites (courtesy US Geological Survey).

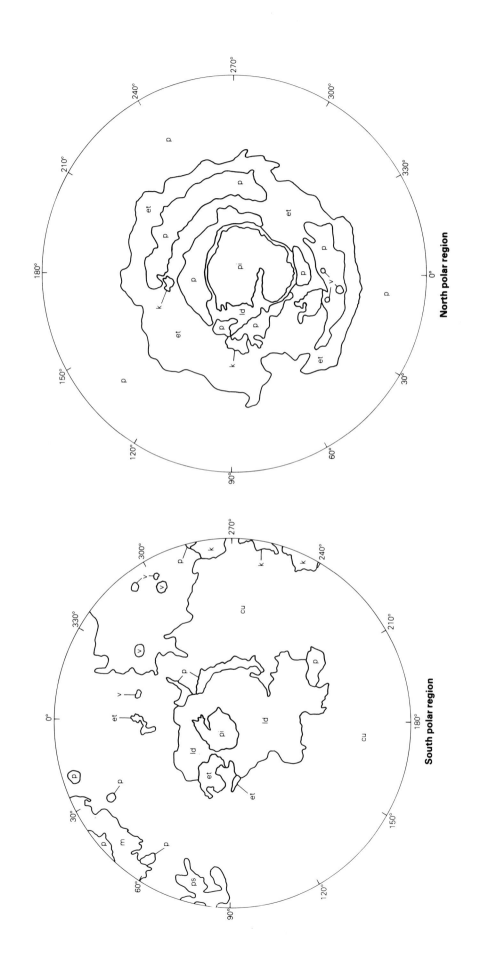

South polar region

North polar region

Figure 7.4 Geological terrain map; massifs (m), heavily cratered plains (pm), ridged plateau plains (pr), volcanic plains (pv), volcanoes (v), ice caps (pi), moderately cratered plains (p), undivided plains (cu), heavily cratered terrain (cu), layered terrain (ld), etched terrain (et), canyonlands (c), chaotic terrain (hc), fretted terrain (hf), knobby terrain (k), and grooved terrain (g) (modified from Scott and Carr, 1978). Numbers indicate the position of impact basins, keyed to Table 7.2.

Figure 7.5 Oblique view obtained by Viking orbiter across the Argyre basin showing the rugged massifs that define the rim. The floor of the basin appears to have been flooded with lavas. Horizon in the background shows haze layers 25–40 km high, thought to be crystals of carbon dioxide (JPL P-17022).

be filled with volcanic plains and is mantled with aeolian deposits. The 900 km diameter Argyre basin (Fig. 7.5) is also in the southern hemisphere and appears to be mantled. Much better images exist of the Isidis basin, a ~ 1100 km feature in the northern hemisphere. The floor of this feature has been buried by flood lavas and, later, larger outpourings of lava occurred to the east of the basin.

All three basins are defined principally by rugged **massifs** or mountain blocks (unit m), 20 km or smaller across and inward-facing scarps. The massifs tend to be closely spaced near the basin and become more widely spaced and subdued away from the rim. They are probably analogous to basin-related massifs on the Moon (Fig. 4.4) and Mercury (Fig. 5.25) and represent uplifted blocks of crust. However, radial ejecta deposits are lacking around most martian basins.

7.3.2 Heavily cratered terrain

Heavily cratered terrain (unit cu) occurs mostly in the southern hemisphere. Detailed photogeologic mapping shows that this unit can be further subdivided based on the degree and style of modification. At first appearance, this terrain is similar to the lunar highlands since it is dominated by large (>20 km), degraded impact craters. Closer inspection shows some important differences in comparison with the Moon. Most of the large craters are relatively shallow; ejecta deposits are degraded or lacking, the rims are degraded and the floors are flat and may have been filled with deposits of various origins.

Furthermore, the large craters are more widely spaced than those on the Moon and are separated by complex intercrater plains. The plains are superposed on most of the craters and cover the crater ejecta deposits. Some crater rims and plains display extensive valley networks (Fig. 7.6) of presumed fluvial origin, while other plains show evidence of aeolian processes; thus, some of the plains are probably sedimentary deposits. Most of the plains within the cratered terrain, however, are thought to be volcanic (Greeley and Spudis 1978*b*), and it is likely that their history reflects a complex interplay of fluvial, aeolian and volcanic processes. Superimposed on the large craters and the intercrater plains is a population of small (<20 km), rela-

Figure 7.6 View of cratered terrain in the southern hemisphere showing valley networks and fracturing. View is 200 km across; north is toward lower left (Viking Orbiter 63A09).

Figure 7.7 Pedestal craters in the northern lowland plains. ▷ Pedestals are considered to be surfaces of ejecta deposits that have been armored with rocks to prevent erosion while the surrounding plains have been lowered by deflation. Many of the small mounds with central craters may be small pedestal craters, rather than volcanic cones, as has been proposed for similar features in other areas of Mars. Area shown is 45 by 65 km (Viking Orbiter 60A53).

tively fresh impact craters, most of which have ejecta patterns suggestive of fluidized flow (Fig. 3.4).

7.3.3 Plains and plateaus

In addition to the intercrater plains described above, numerous other plains occur on Mars. These have been mapped as undivided plains, moderately cratered plains, ridged plateau plains and volcanic plains.

Plains (undifferentiated; unit p) occur principally in the northern lowlands and on the floors of large craters and basins. The northern lowland plains exhibit a wide range of landforms when seen at high resolution, some of which are attributed to periglacial processes; others resemble small volcanic features, but could also be eroded impact craters (Fig. 7.7). Although both Vikings landed in the north (Fig. 7.3), it is not known how representative these sites are for the plains as a whole. Nonetheless, images from the landers provide clues to the processes that have shaped the surface (Fig. 1.4). Most of the rocks are thought to be impact ejecta, although some may be derived from frost-heave or other endogenic processes. Aeolian features, including drifts of windblown sediments and possible ventifacts (wind-sculptured features), are abundant at both sites.

In general, these plains of Mars represent diverse origins and processes of modification. The relative lack of large impact craters indicates the relative youth of the surface. As detailed photogeologic mapping progresses, it is likely that these plains will be subdivided by both type and age.

Moderately cratered plains (unit pm) occur as a plateau centered in the Syria Planum–Sinai Planum area and around Alba Patera. As the unit name implies, impact crater frequencies indicate an intermediate age. Extensive grabens and other fractures show that this unit has been subjected to tectonic deformation, mostly crustal extension (Fig. 7.8).

Ridged plateau plains (unit pr) are found primarily in four areas: within the Isidis basin, as a north–south band

(Lunae Planum) centered at 70°W, and in two areas around the Hellas basin. Their primary characteristic is the presence of widely spaced ridges that are very similar to lunar mare ridges, as described by Lucchitta and Klockenbrink (1981) (Fig. 7.9). The unit is relatively smooth on 100 m resolution images but is peppered with small, fresh craters when seen at higher resolution. Ridged plateau plains are considered to be volcanic flood lavas.

Volcanic plains (unit pv) display flow lobes (Fig. 7.10) and other features diagnostic of lava flows. Volcanic plains occur in association with the two principal volcanic provinces, the Tharsis region and the Elysium region (Fig. 7.3). Source vents for most volcanic plains on Mars remain hidden, similar to the mare flows on the Moon.

Like many volcanic plains, the martian units probably originated from fissure vents which were subsequently buried by their own flows or by flows from other sources. By analogy with Earth, the martian volcanic plains probably involve a combination of flood eruption and "plains" eruptions (Fig. 3.27). Some of the volcanic plains appear to have originated on the flanks of the large shield volcanoes, then flowed beyond the margin of the constructs.

7.3.4 Central volcanoes

Martian central volcanoes (unit v) are characterized by localized source vent(s) and occur as shield volcanoes,

Figure 7.9 Mosaic showing ridged plateau plains east of Mellis Dorsa in the southern hemisphere of Mars. Ridges resemble mare ridges on the Moon. Plains have been fractured and the feature at (A) appears to have been offset laterally. Crater in center is a typical ejecta flow crater showing multiple flow lobes (Viking Orbiter 610A01-3, 608A409).

◁ **Figure 7.8** Shaded airbrush relief chart showing Alba Patera, a unique volcanic structure, and the set of fractures and grabens surrounding the caldera region. Area shown is 1000 by 1500 km (courtesy US Geological Survey).

Table 7.3 Classification of large martian volcanoes.

Name	Location	Type	Area (10⁴km²)	Note
Alba Patera	110°, 40°	Alba	113.0	1
Albor Tholus	210°, 19°	Dome	1.94	2
Amphitrites Patera	299°, −59°	Highland patera	6.55	3
Apollinaris Patera	186°, − 8°	Shield	5.35	4
Arsia Mons	121°, − 9°	Shield	33.3	5
Ascraeus Mons	105°, 11°	Shield	14.0	5
Biblis Patera	124°, 2°	Shield	1.18	6
Ceraunius Tholus	97°, 24°	Dome	1.01	2
Elysium Mons	214°, 25°	Shield	13.1	7
Hadriaca Patera	267°, −30°	Highland patera	9.24	3
Hecates Tholus	210°, 32°	Shield	2.59	6
Jovis Tholus	118°, 18°	Shield	0.29	6
Olympus Mons	133°, 18°	Shield	37.4	5
Pavonis Mons	113°, 0°	Shield	15.0	5
"Tempe" Patera	63°, 44°	Highland patera	1.54	3, 8
Tharsis Tholus	91°, 13°	Dome	1.68	2
Tyrrhena Patera	254°, −22°	Highland patera	3.94	3
Ulysses Patera	122°, 3°	Shield	0.85	6
Uranius Patera	93°, 26°	Shield	2.71	6
Uranius Tholus	97°, 26°	Dome	0.38	2

[1] Unique volcanic structure on the terrestrial planets.
[2] May represent buried older shields or different style of volcanic activity.
[3] Interpreted as ash shields with less lava present.
[4] Oldest lava shield recognized on Mars; occurs on plains/uplands boundary.
[5] Main shields of the Tharsis complex.
[6] Shields that have been partly buried by surrounding plains lavas.
[7] Slightly steeper sides than the Tharsis shields.
[8] Informal name, see Plescia and Saunders (1979).

Figure 7.10 High-resolution image showing flow lobes of presumed volcanic origin (Viking Orbiter 806A60).

dome volcanoes, highland patera and as a unique feature—Alba Patera. In addition are numerous small volcanoes including cones, low shields and possible composite cones which are discussed in Section 7.5.

Shield volcanoes provided the first clear evidence that volcanism was important on Mars. Shield volcanoes (Figs. 7.11 and 7.12) in the Tharsis and Elysium regions are composed of thousands of individual flows, many of which were emplaced through lava tubes and channels, typical of shield-forming flows on Earth (Fig. 3.18). The summit areas are characterized by calderas (Fig. 7.13), most of which are complex and exhibit repeated episodes of eruption, collapse and tectonic modification. Many of the martian shield volcanoes have low slopes on the outer flanks and a steeper summit region which may reflect a change in the style of volcanism.

Dome volcanoes (tholii) are smaller and have steeper slopes (Fig. 7.14) than shield volcanoes. The steeper slopes may be attributed to more viscous lavas, incorporation of pyroclastic deposits with lava flows, lower rates and volumes of effusion, or some combination of these factors, any one of which could result in the development of dome-shaped volcanoes. Many of the features on Mars

first identified as domes are now considered to be partly buried shield volcanoes, in which the steeper summit area remains exposed (Table 7.3). Although rare, some features have been found on Mars (Fig. 7.15) that may be similar to composite cones.

Highland patera, as defined by Jeff Plescia and Steve Saunders (1979) of the Jet Propulsion Laboratory, are characterized by low profiles, radial channels and complex central calderas (Fig. 7.16). Some of the features first identified as patera on Mariner 9 are now seen to be of non-volcanic origins. With the exception of Tempe Patera in the northern hemisphere, all the highland patera (Table 7.3) occur near the Hellas basin. Because these features are about the same age and appear to be situated over ring-fractures associated with Hellas, Peterson (1978) suggested that they may represent early-stage vol-

Figure 7.11 Olympus Mons, one of the prominent shield volcanoes in the Tharsis region, measures 600 km across and is surmounted by a complex summit caldera (see Fig. 7.13). The volcano is surrounded by a prominent scarp several kilometers high. Flows derived from the volcano can be traced onto the surrounding plains at (A). North is toward upper left. (Viking Orbiter 646A28).

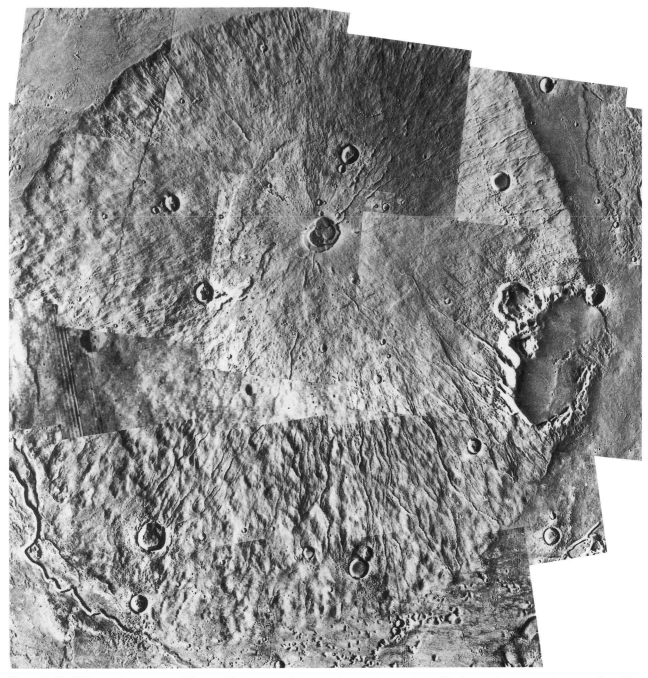

Figure 7.12 Viking orbiter mosaic of Hecates Tholus, one of the prominent volcanoes in the Elysium region, measuring more than 170 km across. Various radial channels have been interpreted as erosional ash channels, lava channels or channels eroded by fluvial processes.

Figure 7.13 Mosaic of high-resolution frames of part of the Olympus Mons summit caldera, showing extensive wrinkle ridges in one of the floor units. These ridges are typical in morphology to those that occur on lunar maria. Area shown is 75 km across (JPL P-19381B).

Figure 7.14 Tharsis Tholus, measuring 110 by 170 km, has steep flanks and is classified as a dome volcano (Viking Orbiter 858A23).

Figure 7.15 This 2 km high feature, located at 21°S, 188°W, may represent a composite cone (Greeley & Spudis 1978b), suggested by slopes 10° to 20° and its rugged slopes. The summit caldera is about 8 km across and appears to be filled with lavas. Numerous channels and possible flows radiate from the caldera. Apron-like flows (arrows) could be either landslides or lava flows (JPL P-24656).

Figure 7.16 Tyrrhena Patera, one of the highland patera associated with the Hellas basin in the southern hemisphere. The area shown is about 280 km across (Viking Orbiter 87A14).

canism associated with post-impact adjustments of the Hellas basin.

Alba Patera represents a volcanic feature that may be unique. Stretching more than 1000 km across, it is the largest central volcano seen anywhere in the Solar System. It is characterized by a set of complex ring fractures surrounding a 100 km caldera centered at 110°W, 40°N (Fig. 7.8). Sheet lavas, channel-fed flows and complex tube-fed lavas can be traced westward from the central caldera for hundreds of kilometers. Small domes that may be either parasitic vents or "rootless" vents occur in association with some of the flows at the western extremity of the volcano.

7.3.5 Polar units

The polar regions incorporate several distinct terrains, including *ice caps* (unit pi), *layered terrain* (unit ld) and *etched terrain* (unit et). The ice caps form bright zones around the north and south poles. They persist through the late spring and summer and, at least in the north, are considered to contain both water ice and frozen CO_2, as inferred from infra-red data obtained from the Viking orbiters (Kieffer and Palluconi 1979; Farmer *et al.* 1977). As the caps retreat with the warm seasons, outlying patches of CO_2 frost often remain in low-lying and shaded areas, such as behind crater rims.

Layered terrain (Fig. 7.17) occurs in both polar areas. In the north it is superposed on plains units and in the south it overlies cratered terrain. Layering is visible as fine bright and dark bands exposed on slopes. Thicknesses of individual bands have been estimated at 10–50 m, but they probably exist as thinner units which are simply too small to be seen. Many of the bands can be traced hundreds of kilometers and most of them tend to follow the topography, suggesting that the bands represent nearly horizontal layers. Some bands transect (cross-cut) underlying sets of bands, indicating discontinuities in deposition. The origin of the layers has evoked considerable controversy, as reviewed by Cutts and Lewis (1982). Most investigators suggest that they are alternating deposits of dust, water ice, CO_2 ice or some combination of materials that accumulate on an annual or longer term cycle. A major paradox is the seemingly young age of the polar layered terrains. The number of layers indicate

Figure 7.17 (top) View of layered terrain in the north polar area. Area shown is 65 km across, centered at 80°N, 347°W (Viking Orbiter 56B84).

Figure 7.18 (above) Etched terrain in the south polar region, centered at 76°S, 74°W, showing characteristic pitted surfaces that may be the result of deflation and/or ablation of ice. Area shown is 200 km across (Viking Orbiter 390B90).

an age <1 my. The number of surrounding pedestal craters (see Section 7.7.1) indicate an age younger than Olympus Mons (P. Schultz, personal communication).

Etched terrain consists of areas that have been eroded to form pits and valleys (Fig. 7.18). First described by Sharp (1973), this unit appears to involve layered terrain that has been eroded by some processes such as wind deflation. For example, ablation of ice layers may leave behind loose grains of sand which can then be removed by the wind.

7.3.6 Canyonlands

Stretching 4000 km along the equatorial zone eastward from the Tharsis Montes are grabens, canyons, pit craters and channels which form vast *canyonlands* (unit c). Its length is the equivalent of a canyon system extending across the United States from the Pacific to the Atlantic Oceans! The region, including the main canyon system, was named Valles Marineris in honor of the Mariner 9 spacecraft which returned the first pictures of this remarkable area. Individual canyons are as wide as 200 km and as deep as 7 km. As shown in Figure 7.19, the canyonlands can be subdivided into three sections, Noctis Labyrinthus in the west, the main section of canyons in the center and a complex eastern section. Noctis Labyrinthus consists of short, narrow canyons that create a mosaic of crustal blocks (Fig. 7.20). The area is slightly east of the *Tharsis bulge,* a prominent "swelling" on Mars which rises more than 11 km above datum. Most of the canyons appear to be grabens that formed in response to extension associated with the uplift of the Tharsis bulge.

The central part of Valles Marineris is more than 2400 km long and consists of multiple, parallel canyons, all trending east–west, with smaller grabens and chains of coalescing pits. The walls of the canyons exhibit layering which could represent the sequences of the lava flows that are thought to make up the plains into which the canyons have cut.

The eastern part of Valles Marineris loses the prominent east–west fabric and consists of loosely connected canyons and depressions, some of which are as large as 150 km across. These grade into enormous channels, most of which eventually empty into Chryse Planitia to the north. The general area is classified as *chaotic terrain* (discussed below) and is characterized by its jumbled, irregular appearance (Fig. 7.21), suggestive of extensive mass wasting.

Most investigators agree that tectonic processes, somehow related to the Tharsis region, played a key role in the origin of the canyonlands. Crustal extension generated fractures and grabens that were enlarged by mass wasting (Fig. 7.22) and further eroded by aeolian and other processes. From analysis of deposits emplaced within some canyons, Jack McCauley (1978) of the US Geological

168

Figure 7.19 Shaded airbrush relief chart showing the Valles Marineris region. This area, known collectively as the canyonlands, is composed of three elements: Noctis Labyrinthus in the west, the central canyon area, and an eastern area made up of chaotic terrain and outflow channels. Area shown is about 3500 by 6000 km (courtesy US Geological Survey, Flagstaff).

Survey proposed that vast lakes were once contained within the canyons. He suggests that as dams within the canyonlands were eroded, the lakes drained eastward through the chaotic terrain.

Groundwater processes and subsurface ice are also involved in most models of the evolution of Valles Marineris. Some canyonland elements, such as Hebes Chasma, are closed systems and ablation of ice to the atmosphere and subsurface erosion by flow of groundwater through "piping" may account for some of the material removed from these areas.

7.3.7 Channeled terrain

"Dry riverbeds were one of the most unexpected finds of the Mariner 9 mission to Mars": so wrote David Pieri in his 1983 review of the valleys and channels on Mars. Why unexpected? Because by the Mariner 9 mission, it was clearly established that liquid water can exist on the surface of Mars for only short periods of time (~ minutes)—yet the features revealed by the Mariner 9 cameras clearly spoke of erosion by water. This implied either that environmental conditions on Mars were quite different in the past or that the features were formed by processes other than free-flowing water.

Two types of features have been identified: *outflow channels* (Fig. 7.21) and *small valleys* (Fig. 7.6). Small valleys and valley networks are discussed in Section 7.7.2. The *channeled terrain* (unit ch) refers to areas that have been modified by outflow channels and includes terrain modified by the "fretted" channels of Sharp and Malin (1975). Outflow channels generally begin in terrain that has been modified by collapse and mass wasting.

Figure 7.20 Mosaic of Viking orbiter images showing Noctis Labyrinthus, the western part of the Valles Marineris canyonlands. Grabens and other fractures resulted from the uplift of the Tharsis bulge (from US Geological Survey MC17 NE).

Most of these channels originate fully developed in uplands and flow northward across diverse terrains onto the lowland plains. They are typically 800–1200 km long, although some, such as Kasei Vallis, may reach 3000 km in length. Some outflow channels are 2.5 km deep and exceed 150 km in width. In general, channel depth is inversely related to width. Gradients for outflow channels have been estimated from a minimum of 0.5 m km⁻¹ for Simud Vallis to a maximum of 20 m km⁻¹ for parts of Maja Vallis.

Many outflow channels exhibit terracing or layering along the walls which may reflect differential erosion, or repeated episodes of channel formation, or both. Episodic channeling is clearly documented in Chryse Planitia by stratigraphic relations, shown in Figure 7.23. Evidence for scouring, grooving and erosional "plucking" are seen in association with many outflow channels which sweep across wide areas (Fig. 7.24).

7.3.8 Modified terrain

Several areas of Mars have been so severely modified that they have been singled out as distinctive terrain units. *Chaotic terrain* (unit hc; Fig. 7.21) is characterized by disordered, jumbled blocks of terrain that appear to have been subjected to extensive mass wasting. In addition to the eastern part of Valles Marineris, chaotic terrain is found along the western boundary of Lunae Planum, south of Apollinaris Patera and north of Hecates Tholus, most of which is on the rings of ancient impact basins.

Fretted terrain (unit hf) is found primarily along the boundary of the cratered terrain and the northern lowland plains. First described by Sharp (1973), it typically consists of two parts, an upper section of relatively undisturbed terrain and a lower section of relatively smooth plains that appears to have formed by scarp retreat at the expense of the upper section (Fig. 7.25). The lower sec-

170

Figure 7.21 Mosaic of Viking orbiter images showing a 20 km wide channel emerging from the chaotic terrain. This channel eventually connects with Simud Vallis and flows northward into Chryse Planitia (JPL P-16983).

Figure 7.22 Oblique Viking orbiter view across Gangis Chasma in the canyonlands. The landslide on the far wall extends as far as 50 km from the canyon wall and is one of several landslides that have enlarged the canyon. Visible in the lower right is a dark deposit which consists of sand dunes, demonstrating aeolian activity (JPL P-16952).

Figure 7.23 Viking orbiter mosaic of a 21 km diameter impact crater in western Chryse Planitia showing multiple channeling episodes. A set of degraded channels (A) is overlain by ejecta deposits from the crater showing that channel formation pre-dates the impact. On the north (upper half of figure) side of the crater the ejecta deposits have been eroded by a later episode of channeling. The regularly spaced dark rings are artifacts of the camera system (from Greeley *et al.* 1977).

Figure 7.24 Part of channel system in Chryse Planitia showing channel segments and teardrop-shaped islands. Area shown is 600 by 750 km, centered at 23°N, 33°W (mosaic part of MC 11-NW, US Geological Survey).

tion often grades imperceptibly into the lowland plains. The "fretting" is clearly controlled by fracture patterns in some areas and by channels in other areas.

A large area northwest of Olympus Mons shield volcano has been extensively modified to produce *grooved terrain* (unit g) which extends as lobes from the base of Olympus Mons several hundred kilometers (Fig. 7.26). The origin of this terrain, known also as the Olympus Mons aureole, has evoked considerable debate and includes models involving erosion of ash sheets (King and Riehle 1974; Morris 1982), eruptions through ice sheets (Hodges and Moore 1979) and mass wasting processes (Harris 1977; Lopes *et al.* 1980 and 1982).

7.3.9 Knobby terrain

Knobby terrain (unit k) is found mostly in the northern hemisphere where it occurs east of Elysium, within the northern plains (Guest *et al.* 1977), and along the cratered uplands–northern lowland plains border where knobs are often gradational with fretted terrain. Individual knobs range up to 10 km across and appear to be of diverse origins, including erosional remnants of older terrain such as highly degraded crater rims (Fig. 7.27). Some knobs stand higher than the upper surfaces of fretted terrain, indicating that these knobs are more than simply erosional

Figure 7.25 View of fretting along the border of the cratered terrain and the northern lowlands. "Fretting" results from scarp retreat and the breakup of the uplands to form small, outlying mesas. Area shown is 250 by 300 km (US Geological Survey, MC5-SC, part).

Figure 7.26 Mosaic showing Olympus Mons and the surrounding aureole material. Northern lobe (A) is bounded on the eastern side by scarp (C) and partially overlain by the northwestern lobe (B) (from Lopes *et al.* 1982; mosaic courtesy of Peggy Thomas).

remnants. In other areas, many of the knobs have small summit craters, suggesting that they may be small volcanoes.

7.4 Craters

When Mariner 4 returned the first spacecraft images in 1964, Mars was seen to be a cratered object similar to the Moon—an impression that did not change substantially with the later flybys of Mariner 6 and 7. As fate would have it, all three spacecraft imaged essentially the same physiographic region of Mars, the heavily cratered terrain. Yet, even though these images were crude by today's standards, the craters appeared to be somewhat different from those on the Moon. Mariner 9, carrying both wide angle and telescopic cameras, revealed not only the diversity of terrains on Mars but also provided additional details on martian craters—details that were further enhanced by Viking images.

7.4.1 Crater morphology

Even on earliest spacecraft images, martian craters were seen to be shallower than lunar craters. This was attributed to erosion of the rims and partial filling of the craters by windblown sediments and other deposits (Fig. 3.35). We know that lunar craters are enlarged by slumping, and if volatiles were present in the "target" materials on Mars, this process would probably be enhanced and could lead to a lower depth-to-diameter ratio.

The morphology of craters might provide some insight into the environment (including "target" properties) at the time the craters were formed. For example, from analysis of Viking orbiter images, Eugene Smith (1976) and other investigators noted that many martian craters have central pits (Fig. 7.28). Wood and his colleagues (1978) attributed this difference to the presence of subsurface ice. They suggested that during impact the transient cavity penetrated a zone containing ice which was volatilized by impact-generated heat and vented to form the central pit.

One of the more striking results of the Viking mission was the discovery of ejecta deposits (Fig. 3.4) which appear to have been emplaced at least partly by flow of fluidized materials (Carr *et al.* 1977). In many cases, the final stages of ejecta emplacement involved ground-hugging flow, rather than ballistic emplacement (Fig. 7.29). These craters, variously termed "splosh" craters, ejecta-flow craters or rampart craters, are not seen on Mercury or the Moon, and it has been suggested that water or the atmosphere of Mars, or both, may have been responsible for the form of the ejecta.

Laboratory impact experiments using viscous targets

Figure 7.27 Knobs east of the Elysium region which represent rim of degraded crater about 280 km across (Viking Orbiter 672A64).

Figure 7.28 Impact crater 50 km in diameter in the Noachis region. The central peak in this crater has a pronounced pit which may have formed by collapse during the final stages of crater formation. The abundance of central pits on Mars suggests that they may be related to volatiles, such as subsurface water on ice contained at the time the impact occurred (Viking orbiter mosaic 211-5735 part).

Figure 7.29 Viking orbiter image of an ejecta-flow crater (upper right). Emplacement of ejecta must have involved material flowing very close to the surface rather than as a ballistically emplaced deposit, indicated by the control of the small impact crater (A). Area shown is 50 km across (Viking Orbiter 32A28).

to simulate impacts on Mars led to a model which accounts for many of the features observed on martian craters (Fig. 7.30). Schultz and Gault (1979) conducted other experiments to assess atmospheric effects on ejecta emplacement and found that ejecta mixed with even a low-density atmosphere could produce distinctive ramparts, flow-like features and radial grooving, depending on the size of the ejecta particles. Thus, it would appear that volatiles in the form of subsurface water (or perhaps ice) or an atmosphere could result in ejecta flow patterns.

Peter Mouginis-Mark (1979; 1981) classified martian craters by their morphology (paying particular attention to ejecta deposits) and searched for correlations with terrain, elevation and latitude. In general, he found that ejecta deposits extend farther from the crater rim at higher latitudes and at lower elevations, perhaps indicating that the fluidity of the ejecta was greater during emplacement. This would suggest that a higher abundance of volatiles was present in those areas at the time of impact.

7.4.2 Crater statistics

Crater counts on Mars have generated considerable controversy. For some areas, the shape of the size-frequency distribution curve is remarkably similar to that of the Moon. Other areas are different from the Moon, depending upon the size range of craters counted. Of course,

it is reasonable to expect some differences between the Moon, where erosion by non-impact processes is negligible, and Mars, where wind and water have modified the surface. Thus, like impact craters on Earth, many craters on Mars may be obliterated or so degraded that they are impossible or difficult to identify for counting. For example, Wood and Head (1976) noted that the Moon appears to have more basin or basin-like craters than Mars. However, more recently, Schultz and co-workers (1982) have found evidence for 21 possible additional basins (Table 7.2). The features they describe have been severely degraded (Fig. 7.31), but many still display geomorphic features indicative of impact.

Furthermore, processes of gradation have not been uniform through time on Mars. Evidence in the old, heavily cratered terrain suggests that running water may have carved drainage networks early in martian history. Younger terrains do not show this evidence, suggesting a change in climate. Thus, the erosion and obliteration of craters may not be taken into account easily as a function of time.

Attempts to relate crater counts to "absolute" ages of surfaces—typically using the cratered surfaces of the Moon that have been sampled and dated as a calibration—are fraught with difficulties (Neukum and Hiller 1981; Chapman and Jones 1977). In general, martian crater counts exhibit more small craters in proportion to large craters than is typical for the Moon. Soderblom and his colleagues (1974) attribute this to a greater proportion of secondary craters on Mars (perhaps resulting from the break-up of ejecta in the martian atmosphere to produce more secondaries); however, Neukum and Wise (1976) suggest it may be due to a fundamental difference in the distribution of impacting bodies in the vicinity of Mars.

Despite these difficulties, crater counts on Mars provide unique and fundamental information for *relative* dates of surfaces that have not been extensively modified.

7.5 Volcanism

The history of the martian surface is dominated by volcanism. A study based principally on the morphology of surface features shows that a substantial part of the surface of Mars involves various volcanic units (Greeley and Spudis 1981). Although the enormous shield volcanoes are impressive, they constitute less than 1 percent of the total surface. In contrast, plains of probable volcanic origin cover more than 35 percent of Mars. Estimates of the composition of volcanic materials on Mars are derived from analyses of surface materials at the Viking lander sites using X-ray fluorescence techniques, from remote-sensing data based on spectral characteristics, and from theoretical modeling based primarily on geophysical and petrological considerations. Together, the results suggest igneous rocks (such as basalts) that are mafic to ultramafic in composition.

7.5.1 Styles of volcanism

Mars exhibits a wide variety of volcanic features, probably reflecting many different styles of volcanism. Some of the vast plains (Fig. 7.9) may have formed by flood eruptions of mafic lavas. Lack of large channels and collapsed lava tubes in these areas may signal rapid emplacement and ponding of thick sheets of very fluid lavas, analogous to some mare units on the Moon and possibly some of the mercurian smooth plains. Other plains exhibit hummocky surfaces and small channels that could be volcanic (Fig. 7.32). These plains may have been emplaced by lavas erupted at lower rates (Fig. 3.27). Finally, some plains, such as those of Syrtis Major, have central vents and may be very low-profile shield volcanoes (Schaber 1982a).

Some plains may not be lava flows but could be vast volcanic ash flows. Water almost certainly has been present throughout most of the history of Mars, if not on the surface then in the subsurface as groundwater or ground ice (Fanale 1976). With widespread effusive volcanism, it is likely that phreatomagmatic eruptions would have generated abundant ash deposits, as Dave Scott (1982) has suggested for the mantling deposits seen in the Memnonia

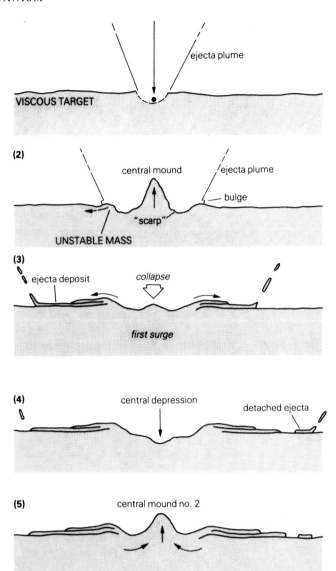

Figure 7.30 Sequence of impact cratering in viscous targets, derived from analysis of laboratory experiments (from Greeley *et al.* 1980).

region (Fig. 7.33). However, Francis and Wood (1982) argue by analogy with terrestrial volcanism that such eruptive styles may not be present to this degree on Mars, and as an alternative, Schultz and Lutz-Garihan (1982) suggest that these thick mantle deposits may be related to ancient polar processes.

From photogeologic analysis, Greeley and Spudis (1981) concluded that the highland patera may be similar

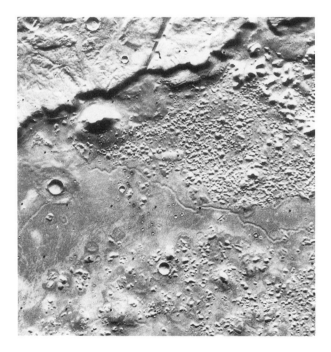

Figure 7.31 (top left) Viking orbiter mosaic showing multi-ring pattern associated with Aram Chaos, centered at 3°N, 22°W. Concentric patterns of unstable chaotic terrain and stable terrains delineate ancient multi-ring basin pattern. Area shown is 1000 by 1000 km (from Schultz *et al.* 1982).

Figure 7.32 (left) Plains in the Cydonia region (41°N, 9°W) showing a sinuous channel that may be analogous to lunar sinuous rilles (Greeley and Spudis 1981). Smooth material flanking the channel and mantling the adjacent terrain could be lava deposits associated with the channel. Area shown is 40 by 40 km (Viking Orbiter 72A11).

Figure 7.33 (above) Postulated ignimbrite deposits that overly ancient cratered terrain in the Memnonia region of Mars, showing possible nonwelded (smooth) and welded (striated) zones (Scott 1982). Area shown is 240 by 290 km (Viking Orbiter 599A53).

to ash shields seen on Earth (Fig. 3.18). We speculated that early-stage eruptions through groundwater contained in impact-generated regolith produced phreatomagmatic deposits. With continued eruption, activity evolved into effusion of lava flows, part of which carved channels in the ash deposits leading to the forms exemplified by Tyrrhena Patera (Fig. 7.16).

Phreatic eruptions are also suggested on a smaller scale. In the northern plains, some of the small cones with summit craters have been suggested by Greeley and Theilig (1978) and Frey *et al.* (1979) to be analogous to "pseudocraters" in Iceland. Pseudocraters are small pyroclastic cones (often with summit craters) which form on the surface of lavas that have flowed over water-saturated ground. Local phreatic explosions rupture the crust of the flow and generate piles of tephra.

Except for their size, martian shield volcanoes mimic Hawaiian shields. Careful mapping and analysis of impact crater frequencies by Crumpler and Aubele (1978) show that Arsia Mons, a typical martian shield volcano, developed in the following sequence: (a) construction of the main shield; (b) parasitic eruptions on the northeast and southwest flanks; (c) caldera formation by subsidence of the summit region, perhaps in response to magma withdrawal associated with the parasitic eruptions; and (d) effusive activity through fissures at the summit and on the flanks of the shield.

Olympus Mons (Fig. 7.11) is younger than Arsia Mons and does not appear to have evolved to the same degree, at least so far as the complexity of the caldera is concerned. Its total volume is estimated to be three times that of the entire Hawaiian Emperor chain of volcanoes on Earth. Carr (1973) attributes this to a longer period of activity for Olympus Mons and to differences in tectonic style, noting that the Hawaiian chain of volcanoes are probably produced as the Pacific plate slides over a magma "hot spot". Mars evidently lacks Earth-like plate tectonics. Hot spots in the martian interior may produce lava continually through a "fixed" central vent. Some investigators speculate that, in order for the magma to produce the great heights of the martian volcanoes, the lithosphere must be relatively thick in the Tharsis region.

7.6 Tectonism

Tectonic features, such as faults, provide clues to the style of crustal deformation and can be used to place constraints on models of planetary interiors. When the timing of tectonic deformation can be determined, further knowledge can be gained about the evolution of the interior. Although some tectonic features have been found in association with the Hellas, Argyre and Isidis basins, by far the dominant tectonic province on Mars is the Tharsis region. Much attention has been focused on Tharsis in an attempt to relate the tectonic and volcanic processes to the nature of the interior of Mars and its thermal history.

7.6.1 The Tharsis region

The Tharsis region is a broad plateau (Fig. 7.34) some 4000 km across which stands more than 10 km above the surrounding plains. It is dominated by central volcanoes and volcanic plains which vary greatly in age. The "apex" of the province is the Tharsis bulge, an enormous swelling centered at about 0° latitude, 114°W.

Numerous faults and fractures cut the plains (Fig. 7.35) and some of the volcanoes. In general, these are extensional features that are radial and subradial to the Tharsis bulge. Fractures, faults and grabens, some as long as 3000 km, are particularly well developed in three areas: north of Tharsis around Alba Patera, northeast of Tharsis in the Tempe region, and south of Tharsis in the Claritas region. Extensive mare-type ridges in Lunae Planum, Sinai Planum and Solis Planum are crudely concentric to the bulge and may reflect compressive forces. As discussed earlier, part of the canyonlands (Noctis Labyrinthus) is also associated with the Tharsis bulge. In all, the Tharsis region and its associated tectonic features leave their imprint on more than a third of the martian surface.

By precise tracking of the orbital paths of the Mariner 9 and Viking spacecrafts, gravity data have been derived for parts of Mars. These data show that a large, positive, free-air gravity anomaly is centered over the Tharsis bulge (Sjogren et al. 1975), and that local gravity highs occur over the large volcanoes (Sjogren 1979). On Earth, free-air gravity anomalies are typically small because mountains are compensated by low-density "roots". The anomalies on Mars are much larger than on Earth, suggesting a different mode of support, such as mantle convection or a very thick lithosphere.

7.6.2 History and origin of the Tharsis region

Three lines of evidence are used to deduce the history of the Tharsis region (Carr 1981): (a) analysis of the intersections of various sets of fractures; (b) relative age dating of surfaces based on crater counts; and (c) study of the geometric relationships between the fractures and the dated surfaces. Using this process, Don Wise and his co-workers (1979) found that the oldest fractures occur in the cratered terrain of the Tempe region and inferred that uplift of the Tharsis bulge began early in the history of Mars. Carr speculates that the origin of the bulge may be linked to mantle convection associated with the separation of the core. A second, later episode of fracturing occurred in the northern Tharsis area, forming the Ceraunius Fossae and the extensive faults around Alba Patera (Fig. 7.8). Most of the volcanic units in the center of the Tharsis region were emplaced after the main episodes of fracturing. Detailed photogeologic mapping by Scott and Tanaka (1981) led to the derivation of paleostratigraphic maps which show the evolution of the region. They note that the emplacement of volcanic plains and the formation of central volcanoes occurred concurrently throughout the history of Tharsis.

Sean Solomon (1981) reviewed the various geophysical models that have been proposed to explain the Tharsis bulge and its related surface features. He notes that the

Figure 7.34 Shaded airbrush relief map of the Tharsis region, centered at 7°N, 113°W (courtesy US Geological Survey, Flagstaff).

"traditional" model involves broad updoming of the martian lithosphere by a thermal or chemical anomaly, leading to extensional fracturing. Calculations of the stresses in the martian lithosphere required by this model, however, raise some serious questions. Phillips and Lambeck (1980) note that the Tharsis bulge is so large in relation to the planetary radius that full spherical calculations must be made in deriving geophysical models. When such calculations are carried out, the stress geometry for the "traditional" model does not fit the fracture patterns seen on the surface. Consequently, two alternative models have been proposed: (a) Sleep and Phillips (1979) suggest an isostatic model involving a low-density region of the mantle several hundred kilometers deep which may be partly thermal in origin and (b) a *flexural model* proposed by Solomon and Head (1982a) in which the bulge is

largely the result of the accumulation of lava flows supported by a thick lithosphere.

Solomon suggests that some combination of these models may be the best explanation. He notes that older fault systems are best explained by the isostatic model, whereas the younger volcanic terrains fit best with the flexural model. Thus, there may have been a change in the evolution of the interior of Mars from a dynamic mantle to a more static mantle overlain by a relatively thick lithosphere.

7.7 Gradation

Gradation involves the weathering, erosion, transportation and deposition of materials on planetary surfaces.

Figure 7.35 Viking orbiter mosaic showing Ceraunius Fossae of the Tharsis region and two of the smaller volcanoes, Uranius Tholus (upper) and Ceraunius Tholus (100 km in diameter; lower) (Viking Orbiter 211-5639 part).

The present environment on Mars is cold and involves a low-density, CO_2 atmosphere (Table 2.1), and chemical and physical weathering processes probably operate slowly compared to Earth. Nonetheless, views of the martian surface taken from the Viking landers (Figs. 1.1 and 1.4) and from orbit show evidence for gradation by aeolian activity, running water, mass wasting and periglacial processes.

7.7.1 Wind

The low-density atmosphere requires high wind speeds for the movement of particles on Mars (Fig. 3.36). Measurements of winds from the Viking landers and observations of dust storms show that threshold velocities are achieved on Mars and that aeolian processes play a key role in the present-day modification of the surface.

Wind erosion and deflation have led to the formation of wind-deflated areas in polar regions (Fig. 7.18), yardangs (wind sculptured hills; Ward 1979) and exhumed terrain (Fig. 7.36). One class of impact crater, termed *pedestal craters* (Fig. 7.7), appears to have formed on plains that were subsequently eroded, perhaps by the wind. In some cases, the cratered plains may have been first mantled and then deflated. In either case, it has been suggested that blocky crater ejecta "armored" the surface to protect it from deflation so that it remained as a high-standing pedestal while the surrounding surface was lowered.

Since their discovery on Mariner 9 images, sand dunes have been found in many areas on Mars. The largest field occurs in the north polar region (Fig. 7.37) and is equal in size to large sand seas on Earth (Tsoar *et al.* 1979). Smaller dune fields are found elsewhere on Mars (Fig. 7.38), often in association with craters (Fig. 7.39). In general, martian dunes exhibit the same morphology as terrestrial dunes, although star dunes and longitudinal dunes appear to be lacking.

Wind streaks are the most common type of aeolian feature on Mars. They can be either bright or dark relative to the surrounding surface and occur in a wide variety of geometries (Figs. 7.40 through 7.42). Because some change their size and shape with time, wind streaks are also termed *variable features* (Sagan *et al.* 1972). They are considered to result from wind erosion and deposition and are found in many regions where they can be used as local "wind vanes". For example, Peter Thomas and Joseph Veverka (1979) of Cornell University have mapped wind streaks to determine patterns of atmospheric circulation, following earlier work by Sagan *et al.* (1973).

7.7.2 Water

Depending upon the method of analysis, the total amount of water evolved on Mars would evenly cover the surface from 10 to 160 m. Although liquid water cannot exist on the surface today, various channels and small valleys are generally regarded to be the result of flowing water. Because of the critical implications held by the channels

Figure 7.36 Region southeast of Elysium showing various aeolian features. Cratered terrain in the lower part of picture appears to have been mantled by smooth deposits of possible aeolian origin some of which have subsequently been eroded; half of the crater at (A) has been exhumed. The feature above appears to be a crater buried by the mantling units. Crater at (B) post-dates the mantling deposit, as its ejecta is superimposed on the mantle. Various linear features in the lower part of the picture may be wind-sculptured hills, or yardangs. Pattern in the upper left of picture may be sand dunes composed of weathered particles derived from the mantling deposit. Area shown is approximately 80 by 80 km (Viking Orbiter frame 438S01).

Figure 7.37 Compound barchan dunes in the north polar area. Area shown is 40 by 45 km (Viking Orbiter 544B07).

182

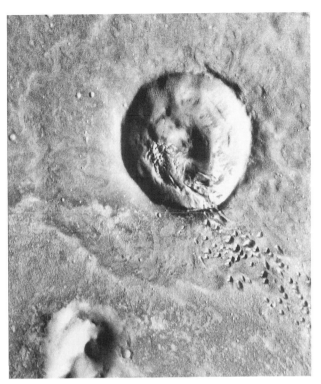

Figure 7.38 Dunes in the cratered terrain showing a series of transverse dunes in the upper left and small barchan dunes in the right half of the picture. Area shown is 50 km across (Viking Orbiter 575B60).

Figure 7.39 "Climbing" dunes which appear to be drifting up and over the rim of a 16 km crater (Viking Orbiter 571B53).

Figure 7.40 Wind streaks. (a, left) Dark wind streaks associated with craters; dark streaks are thought to result from erosional processes. Area is 35 km across (Viking Orbiter 56A20). (b, right) Bright wind streaks associated with craters; most bright streaks are thought to result from depositional processes. Area shown is 35 km across (Viking Orbiter 45B46).

Figure 7.41 Bright and dark streaks associated with craters, suggesting two modes of origin and two winds of opposite directions. Area shown is 200 by 200 km (Viking Orbiter 553A54).

Figure 7.42 Various wind streaks associated with small hills and craters (Viking Orbiter 883A3).

for the history of Mars, NASA formed a working group of scientists to study the topic. Their findings, presented by the Mars Channel Working Group (MCWG 1983), provide an excellent summary of both the outflow channels and the small valleys.

The origin of the outflow channels (Fig. 7.21) remains a controversial topic. A wide range of fluids have been proposed as erosional agents, including running water, ice, lava (Schonfeld 1977), wind (Cutts and Blasius 1981) and mud (Nummedal 1978). Ideas accepted by most students of Mars involve water released by catastrophic floods, an idea first proposed by Hal Masursky (1973) who suggested that the channeled scablands of the Pacific Northwest were a good analog (Fig. 7.43). The scablands formed during the Pleistocene when ice dams broke to release enormous volumes of water that stripped away the basaltic plains. Baker and Nummedal (1978) and Baker (1982) expanded this idea and developed various geomorphologic arguments for the similarities between terrestrial scablands and martian features, although they noted that one to two orders of magnitude greater flow would be required to form the outflow channels on Mars.

Several explanations have been offered for the sudden release of water on Mars. Masursky and colleagues (1977) proposed releases of water as a consequence of ground ice melting, associated with volcanic activity. Scarp re-

treat was suggested by Soderblom and Wenner (1978) to tap subsurface reservoirs of water. Schultz *et al.* (1982) have noted that many of the outflow channels begin on the rings of heavily degraded impact basins, thereby leading to the suggestion that magma intruded along ancient impact-generated fractures and hydrothermally melted trapped volatiles. Mike Carr (1979) proposed an elegant model for ground collapse and fluid release resulting from outbursts of water from confined aquifers. He estimated the required pore pressures on Mars and concluded that flow could be sustained for long periods, even in today's environment.

Small valleys and valley networks (Fig. 7.6) appear to lack the various features indicative of the types of flow seen in the outflow channels, though the features may simply be too small to be seen on available images. The valleys range in length from a few kilometers to nearly 1000 km, although most are only tens of kilometers long and are typically about 1 km wide. Large, undissected upland surfaces occur between the valleys, suggesting immature development.

The valleys often form networks which superficially resemble dendritic drainage patterns on Earth. Careful analysis of the junction angles by David Pieri (1979) shows a much narrower range than is typical for Earth, suggesting a different mode of development from the simple run-off of surface water. Furthermore, the valleys often begin and end for no apparent reason (similar to streams in karst regions), although their courses appear to be controlled locally by topography.

Unlike the outflow channels which cut across diverse terrain units, small martian valleys are restricted almost entirely to the ancient cratered terrain. Their origin has been ascribed to both groundwater sapping processes and to run-off of surface water, perhaps associated with rain. Because they are not found in younger terrain, some investigators have suggested that there was a change in the environment which prohibited their formation.

Particularly thought-provoking is the possibility that large standing bodies of water existed at one time on Mars. Many of the craters in the southern highlands have channels which appear to have emptied onto their floors. The floors of craters such as Gusev are very smooth and have been interpreted as a type of martian playa, or dry lake bed (Squyres, 1989). In addition, observations of possible shoreline features in the northern plains have led some investigators to consider more extensive bodies of water, perhaps even oceans. For example, Vic Baker and his colleagues at the University of Arizona have proposed that much of the northern hemisphere was covered at one time by water.

7.7.3 Mass wasting

Mass wasting of several styles has occurred on Mars, as reviewed by Baerbal Lucchitta (1979; 1981) of the U.S. Geological Survey. Vast landslides have enlarged segments of the canyonlands (Fig. 7.22) and features resembling rock glaciers have been found within the fretted terrain (Fig. 7.44). As discussed by Squyres (1979), many areas, particularly along the boundary between the cratered terrain and the northern lowland plains, exhibit mesas and hills surrounded by debris aprons (Fig. 7.45). The *aureole deposits* (unit g) around Olympus Mons (Fig. 7.26) and similar terrain west of Arsia Mons are ascribed to mass wasting. The precise mode of mass wasting, however, has not been determined; the rate of movement

Figure 7.43 (top left) Landsat image of part of the Columbia Plateau in the Pacific Northwest. Dark areas are enormous channels that resulted from the breakup of an ice dam and flood during the Pleistocene. Flood waters stripped away soil and eroded the underlying basalt flows. Area shown is 150 by 200 km (Landsat frame E-5 854-16504-S).

Figure 7.44 (opposite) View of the Nilosyrtis area at 34°N, 290°W. Ridges and grooves on the material within the valleys may indicate flow of material away from the valley walls and down the valley toward the bottom of the picture. Area shown is about 35 by 40 km (Viking Orbiter frame 884A73).

Figure 7.45 View of debris flows in cratered terrain east of Hellas. Some of the debris aprons extend 20 km from the source. Area shown is 220 by 270 km. North is toward upper left (Viking Orbiter 97A62).

and the amount of water present when the movement occurred are unknown. Although the surface areas of most mass wasting features are too small to date by crater frequency distributions, crater counts on some of the larger landslides suggest that they are relatively ancient.

7.7.4 Glacial and periglacial features

Rossbacher and Judson (1981) calculate that the thickness of the **cryosphere** (subsurface zone where water would freeze) on Mars ranges from 3.0 km thick at the north pole to 1.1 km thick at the equator, taking into account the surface temperature and estimates of heat flow. They note that "sinks" for water on Mars include the atmosphere, polar caps and surface frost, exospheric escape and subsurface (or ground) ice. Even using conservative estimates, they conclude that 90 percent of the total volume of water evolved on Mars could be found today as ground ice. Clifford and Hillel (1983), however, argue that ground ice could not exist over martian geologic history; some major recharging mechanism is necessary. In any case, there is opportunity for the formation of various surface features associated with ice.

Several investigators have speculated that glaciers may have existed on Mars. Lucchitta (1982) examined the outflow channels and compared their size and morphology to large, ice-sculptured valleys on Earth. She found that the features characteristic of glaciated terrain (anastomosing valleys, U-shaped valley cross-sections, hanging valleys and various linear scour features) can all be identified

with many of the martian channels. Lucchitta suggested that at least some of the martian outflow channels could owe their origin and evolution to both fluvial and glacial processes.

As discussed earlier, some small volcanic features may have formed by the interaction of magma and ice on Mars. Hodges and Moore (1979) have carried this possibility to a larger scale and speculated that the scarp around Olympus Mons and its aureole deposits (Fig. 7.26) resulted from subglacial eruptions. They proposed that an ice sheet several kilometers thick may have existed in the Tharsis region. Such an ice sheet may have been related to different locations of the martian polar regions. Peter Schultz (personal communication) argues that polar wandering occurred early in martian history. His evidence includes the existence of thick mantled and layered terrains antipodal to each other and active erosional processes very similar to processes presently acting on and around the present polar deposits. Polar wandering (re-orientation of the spin axis) could account for narrow valley networks occurring within 15° of the present south pole and the distribution of many of the periglacial features found on Mars at a wide range of latitudes.

Numerous other landforms have been suggested to result from ground ice, as reviewed by Lucchitta (1982). Theilig and Greeley (1979) proposed that irregular depressions found in the northern plains (Fig. 7.46) are analogous with terrestrial *alases*, or collapse pits which form from the freezing and thawing of ground ice. Patterned ground of several types also occurs in the northern plains, first described by Carr and Schaber (1977). Poly-

Figure 7.46 View of plateau-forming material in Chryse Planitia showing irregular depressions (A) similar in size and form to *alases* which develop on Earth by the collapse of ground containing ice and water. The scalloped edge (B) of the plateau may indicate scarp retreat enhanced by coalescing depressions. Impact craters (C) are clearly distinguished by their raised rims and circular outline. Area shown is 35 by 35 km (Viking Orbiter 8A74).

Figure 7.47 Patterned ground in northern plains at 44°N, 18°W; although resembling ice-wedge polygons on Earth, which may be as large as 100 m across, these martian features, 5–10 km across, are orders of magnitude larger (Viking Orbiter 32A18).

gons (Fig. 7.47) average 5–10 km across and have been proposed to be ice-wedge features similar to polygonal ground on Earth (Fig. 3.44), tectonic features (Pechmann 1980) or patterns produced by cooling lavas (Morris and Underwood 1978). Equally enigmatic are various curvilinear features (Fig. 7.48) which could be solifluction lobes, linear ice-cored ridges or the result of scarp retreat, as reviewed by Rossbacher and Judson (1981).

7.8 Geologic history

The martian surface displays a rich history of evolution that is only just now beginning to be understood. As demonstrated in Figure 7.49, analysis of landforms and general photogeology enable processes and sequences of formation to be assessed. The early history of Mars is considered to be similar to that of the Moon. Planetary accretion and the period of heavy impact bombardment were accompanied by outgassing and global melting, leading to the formation of a differentiated crust; the earliest discernible history is indicated by remnants of impact basins. Deep-seated structural weaknesses produced by these basins would later control subsequent volcanic activity in a manner much like that on the Moon, but complicated by the presence of ice and the longer-lasting thermal engine of Mars. In the final stages of heavy bombardment, volcanism produced some of the plains in the cratered uplands. By the decline of heavy impact cratering, the terrain dichotomy between the northern and southern hemispheres was established and initial uplift of the Tharsis bulge resulted in crustal fracturing. The reason for the terrain dichotomy remains unsolved in understanding the evolution of Mars.

Outgassing associated with early volcanism and impact bombardment probably generated an atmosphere that was much denser than that found on Mars today. This may have allowed liquid water on the surface and could account for the small valleys which are pervasive throughout the ancient terrains on Mars. This stage may not have lasted very long; the small valleys are not formed in younger terrains and the network patterns are in an immature stage of development.

A change in climate and/or a reduction in the rate of outgassing evidently led to an environment prohibiting surface water. Regolith generated by heavy impact cratering would have been an excellent aquifer and the water responsible for the formation of the valleys may have percolated into the subsurface to form an extensive groundwater system.

Volcanic eruptions, which appear to have operated throughout most of the history of Mars, may have interacted with groundwater to produce volcanic ash plains and patera, the earliest central volcanoes. Volcanism in-

volving lava flows erupted from central vents led to the formation of Alba Patera, the domes and shields in the Tharsis region, and the volcanoes in the Elysium region. Eruptions of very fluid mafic and ultramafic lavas, presumably from fissures, flooded the northern plains.

Catastrophic release of subsurface water appears to have been responsible for the formation of the outflow channels. Although generally younger than the small valleys, these features formed over a wide span of martian history, with the same channels being re-occupied repeatedly.

Tectonic processes continued to deform the Tharsis region, as evidenced by fractures which transect intermediate-age plains. Extension associated with the Tharsis bulge opened the canyonlands, which were subsequently enlarged by various processes including aeolian activity and mass wasting.

In later stages of martian history, volcanism focused in the Tharsis and Elysium regions. Aeolian processes have evidently operated throughout the evolution of the surface and their intensity is probably directly proportional to atmospheric density. These processes and polar-related events are currently active on Mars.

Figure 7.48 Curvilinear features in the northern plains which may be the result of periglacial processes. Area shown is 100 by 100 km (Viking Orbiter 11B01).

Figure 7.49 Mosaic of Viking orbiter frames showing a northwestern part of the Chryse region which displays the rich variety of geologic processes that have shaped the surface of Mars. Ridged plains in the top center of the picture are similar to lunar mare plains and may represent flood lavas. Impact craters ranging in size from 100 km to the limit of resolution, 180 m, are visible throughout the area. Larger, older craters have been severely degraded. Tectonic processes have caused fracturing of the ridged plains in the upper part of the picture. An enormous channel sweeps through the area in the lower part of the picture and, along with wind streaks, is evidence for processes of gradation. North is toward the top (Viking Orbiters mosaic by Peggy Thomas).

8 The Jupiter system

8.1 Introduction

Voyager 1's dramatic 1979 flyby of Jupiter and its satellites doubled the number of objects available for geologic analysis. Prior to that time, only the surfaces of Earth, Moon, Mercury and Mars were known. Voyager, for the first time, revealed the complexity and diversity of Jupiter's major satellites and provided tantalizing clues to their surface histories. The features seen on the satellites prompted a re-examination of traditional geologic processes. The new data caused planetologists to consider such bizarre processes as volcanism driven by sulfur compounds, fracturing of ice-rich planetary crusts, and flooding of surfaces by liquid water.

Summaries of knowledge prior to the Pioneer and Voyager missions are provided in a collection of papers (Burns 1977), based on presentations from an international meeting on planetary satellites held in 1974, and in reviews by Johnson (1978) and Morrison (1982a). Pioneers 10 and 11 arrived at Jupiter in 1973 and 1974 and were the first spacecraft sent to the outer Solar System. Although primarily designed to assess the potential hazards of this "new territory", the Pioneers gathered data on small particles, assessed magnetic fields and obtained low-resolution images of Jupiter and the Galilean satellites.

The flybys of Voyagers 1 and 2 in 1979 and 1980 returned an unparalleled wealth of data on Jupiter and its major satellites (Table 8.1). Among the many discoveries provided by Voyager are the first active volcanoes (other than on Earth) in the Solar System and the ring system around Jupiter. Results are described in various journal special issues (Table 1.2), NASA documents (Table 1.3), collections of papers prepared for special meetings (Morrison 1982a; Burns and Mathews 1986) and in reviews by Soderblom (1980), Morrison and Samz (1980) and Morrison (1983).

Pre-space age studies of the Jupiter system began with the discovery of its major satellites (Io, Europa, Ganymede and Callisto) by Galileo Galilei in the early 1600s, using the newly invented telescope. Although he called them the "Medician stars" in honor of his patron, Cosimo de' Medici, they are now generally called the Galilean satellites (although in Italy they are still often referred to as the Medician satellites). The 1950s saw the application of modern photometry, polarimetry and spectrophotometry to assess the colors and albedos of the satellites, while near-infra-red photometry provided clues to the surface compositions. In 1957, Gerard Kuiper suggested that the surfaces of Europa and Ganymede were covered with water ice, a suggestion that was later confirmed by Pilcher and colleagues (1972) who utilized spectroscopic observations.

In the early 1970s, John Lewis (1971) published a theoretical analysis of the interior structure of the satellites. Based on models of Solar System elemental abundances and estimates of the satellite masses, Lewis predicted that Ganymede, Callisto and possibly Europa have rocky cores and mantle/crusts composed predominantly of water ice, while Io is composed mostly of rocky materials (Fig. 8.1). From considerations of the physical properties of ice-rich crusts, Johnson and McGetchin (1973) predicted that topographic features, such as craters, would deform viscously and become "flattened".

Jupiter and its 16 known satellites have often been described as a miniature Solar System. Jupiter emits energy—mimicking a star—and the Galilean satellites appear to be differentiated with respect to their distance from Jupiter. Had it grown more massive, Jupiter quite probably would have become a star. The four innermost moons are very small, apparently rocky bodies. The largest of these four is Amalthea which measures 270 by 165 by 150 km and was imaged by Voyager (Fig. 8.2). As described by Thomas and Veverka (1982), Amalthea is heavily cratered; the largest crater, Pan, is about 90 km across. In addition, there are various grooves and ridges that are tens of kilometers long. The surface of Amalthea is very dark and red, perhaps colored by sulfur emitted from Io, although bright spots also occur in several areas.

The Galilean satellites constitute the next group of four moons and are the primary subject of this chapter. The orbital and rotational geometries of the four, plus Amalthea, is such that the same side of the satellite always points toward Jupiter, leading to the terms **subjove** (side facing Jupiter), **anti-jove, leading hemisphere** (side facing the direction of orbit) and **trailing hemisphere.** The outer eight satellites range in size from ~ 10 to 180 km in diameter, but little is known about their surfaces (Cruikshank *et al.* (1982) provide a review).

Table 8.1 Missions to the outer planets.

Spacecraft	Encounter date	Mission	Event
Pioneer 10	3 Dec. 1973	Jupiter flyby	Investigation of the interplanetary medium, the asteroid belt, and the exploration of Jupiter and its environment. Closest approach to Jupiter 130,000 km. Exited Solar System 14 June 1983; still active.
Pioneer 11	2 Dec. 1974	Jupiter flyby	Information beyond the orbit of Mars; investigation of the interplanetary medium; investigation of asteroid belt; exploration of Jupiter and its environment. Closest approach to Jupiter 34,000 km.
Voyager 1	5 Mar. 1979	Jupiter flyby	1408 images of jovian satellites were obtained. Major discoveries were active volcanism on Io and a ring around Jupiter.
Voyager 2	9 July 1979	Jupiter flyby	1364 images of jovian satellites were obtained revealing high resolution of Jupiter's geologically diverse icy surface satellites, Europa and Ganymede.
Pioneer 11	1 Sept. 1979	Saturn flyby	Transmitted low-resolution images and other data about Saturn, discovered additional rings and moons not previously known.
Voyager 1	13 Nov. 1980	Saturn flyby	900 images of the saturnian satellites were obtained. Major discoveries were the unexpected complexity of the rings and the true nature of Titan's atmosphere.
Voyager 2	26 Aug. 1981	Saturn flyby	1150 images of the saturnian satellites were received. Despite a scan platform malfunction, virtually all important objectives were met.
Voyager 2	24 Jan. 1986	Uranus flyby	7000 images of Uranus, its rings, and its satellites.
Voyager 2	24 Aug. 1989	Neptune flyby	9000 images of Neptune, its rings, and its satellites.

8.2 Io

Earth-based observations have long shown that Io is more red than any other object in the Solar System, including Mars. Possible surface compositions were considered, and it was found that the overall spectrum of Io closely matches that of sulfur (Nelson *et al.* 1982). However, it is also known that the exact color of sulfur is strongly dependent on its temperature and that the presence of even slight quantities of other materials can drastically alter its color. Thus, the interpretation of the surface composition of Io was (and still is) open to questions (Sill and Clark 1982).

Io is about the same size and density as Earth's Moon, suggesting a predominantly silicate composition. Prior to the Voyager encounters, it was reasoned that Io ought to be a cratered object similar to Mars. Not only should its surface be capable of preserving craters, but given the proximity to the asteroid belt and the potential gravitational focusing effect of Jupiter, Io should be rather heavily cratered. The early, low-resolution pictures of Io obtained by Voyager 1 as it approached Io (Fig. 8.3) showed numerous dark, circular features which were initially interpreted as impact craters.

However, as the spacecraft neared Io and returned progressively higher resolution images, the dark features were seen to be quite irregular in form and to lack the features diagnostic of impact craters. Careful and detailed analysis of even the highest resolution images ($\sim$ 1 km) has failed to reveal structures that could be confidently attributed to an impact origin. Thus, for the first time in Solar System exploration, a planetary object was found which either had not experienced impact cratering—an idea immediately dismissed because of the heavily cratered neighbors of Io (Ganymede and Callisto)—or one whose surface was younger than the rate of formation for impact craters large enough to be seen.

A short time after the discovery of Io's lack of impact craters, another discovery and milestone in Solar System exploration was made—a finding which also explained the apparent absence of impact craters. Analysis of navigation images by Morabito and her colleagues (1979) of the Jet Propulsion Laboratory showed a peculiar bright pattern emanating from the surface of Io (Fig. 8.4). Upon closer inspection, this pattern was found to be an enormous volcanic eruption plume which was showering materials over the surface of Io at a prodigious rate and which was capable of burying not only impact craters but other features as well.

This remarkable discovery was, in fact, predicted. Stan Peale, Pat Cassen and Ray Reynolds (1979) had assessed the tidal stresses generated within Io as a consequence of being "pushed–pulled" between the gravitational fields

Figure 8.1 Diagram showing the interiors of the Galilean satellites. Io and Europa are both thought to be composed largely of silicate material. Both Ganymede and Callisto may have lithospheres of ice and cores of silicates with mantles composed of undifferentiated materials. Ganymede, being more massive and containing a larger component of silicates, probably began internal differentiation earlier and continued to undergo segregation after the period during which Callisto underwent differentiation. Scale bars indicate the diameters of the Moon and Mercury.

Figure 8.2 Voyager 1 image of Amalthea; this innermost, rocky satellite of the Jupiter system is about 155 by 270 km (Voyager 1 image 1097J1-001).

of Jupiter and Europa. They calculated the possible heat generated by tidal stresses to be in the order of 10^{13} watts—more than two to three orders of magnitude greater than heat released from normal radioactive decay. This led them to suggest that active volcanoes might be found and they published their suggestions just before the Voyager flyby!

Subsequent to the prediction and the Voyager discovery of active volcanism, Matson and colleagues (1981) studied Earth-based thermal infra-red data and estimated Io's heat flow. They obtained a value about 10 times greater than the amount based solely on tidal heating, suggesting additional heat sources, such as radioactive decay. This value is some 30 times higher than Earth's heat flow. Thus, Io is the most active object in the Solar System explored thus far, an observation borne out by the satellite's myriad volcanic surface features.

8.2.1 Physiography

About 35 percent of the surface of Io was photographed through the combined coverage of Voyagers 1 and 2 at resolutions sufficient to resolve features larger than ~ 5 km. Most of this coverage is in the equatorial and south polar regions (Fig. 8.5). The highest-resolution images (about 0.5 km per pixel) occur in an equatorial zone between 275°W to 360°W longitude.

Preliminary photogeologic mapping by Schaber (1980; and 1982b) shows that Io can be subdivided into three principal units: various plains, vent-related materials and mountain materials. The plains and vent materials are attributed to volcanic processes. The mountains may be crustal materials not necessarily related directly to volcanism. In addition, various possible erosive features and tectonic structures have been identified and mapped. It

Figure 8.3 View taken from Voyager 1 of Io showing the albedo variations and numerous dark spots. The dark spots were initially thought to be impact scars but higher-resolution images revealed them to be volcanic vents. Diffuse, circular zones (arrows) were found to be active volcanic plumes, as shown in Figure 8.4 (Voyager 1 42J1+000).

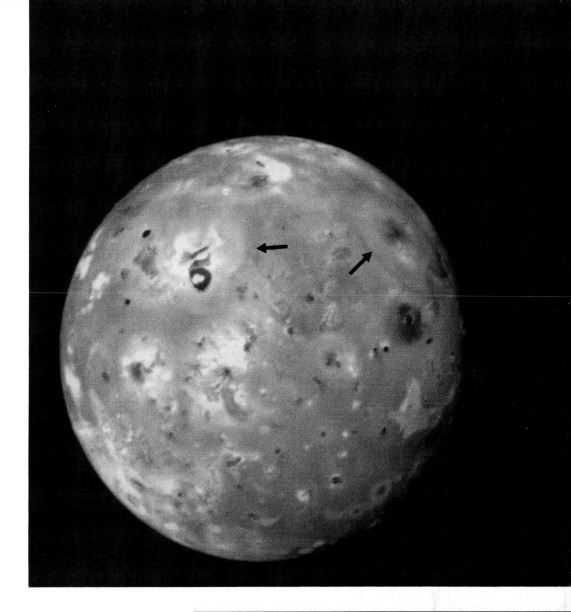

should be noted that the lack of adequate stereoscopic images and impact craters inhibit the determination of stratigraphic relationships. However, embayment and cross-cutting geometries enable some sequences to be determined, at least locally.

In general, the surface of Io displays a wide variety of colors, including various shades of red, yellow, orange and brown, plus very dark and very light areas. Many of these colors can be attributed to sulfur (Sagan 1979) or anhydrous mixtures of sulfur allotropes plus SO_2 frost and sulfurous salts of sodium and potassium (Fanale *et al.* 1979). The controversial topic of whether these possible compositions are representative of the topographic features (such as flows), or simply represent a thin mantling layer (a few millimeters thick), remains open to speculation.

Mountain material Mountain material forms high-standing blocks of rugged relief. These masses can exceed

Figure 8.4 Image showing volcanic plume associated with the vent, Prometheus, as seen in this limb view of Io. The plume extends about 100 km above the surface and forms an umbrella-shaped pattern 300 km wide (Voyager 1 1637748).

Figure 8.5 Shaded airbrush relief map of Io showing various terrains and selected named features. Features in italics refer to active volcanic plumes (courtesy US Geological Survey).

Figure 8.6 High-resolution image of mountainous terrain on Io. This massif, Haemus Mons, is about 200 km across at the base and is estimated to stand more than 9 km above the surrounding plains. Bright halo surrounding the base of the mountain may be fumarolic material (Voyager 1 157J1+000).

Figure 8.7 Voyager image showing layered plains near Lerna Regio, Io. Layered plains are considered to consist of lava flows with interleaved pyroclastic materials derived from volcanic plumes. Bright, wispy zone at the base of the scarp (arrow) is thought to be sulfur dioxide frost, perhaps generated by geyser-like activity; area shown is 880 km wide (Voyager 1 147J1+000).

100 km across and rise more than 9 km above the surrounding surface (Fig. 8.6). Within the area seen at high resolution, mountain material appears to be evenly distributed in regard to both latitude and longitude. Embayment relationships of surrounding plains suggest that mountains comprise the oldest material on Io. In some cases, though, mountains appear to be associated with recently active vents.

Clow and Carr (1980) have analyzed the physical properties of sulfur and demonstrated that features such as steep scarps that are higher than about 1000 m could not be self-supporting. From this analysis, they conclude that the mountains must be composed of something other than sulfur, most probably silicate materials.

Plains   Plains of several types constitute the most widespread units identified on Io. In general, plains range from black and white to various reds and yellows, although plains units in the polar areas tend to be dark brown to black (Masursky *et al.* 1979). *Intervent plains* constitute about 40 percent of the area mapped. These plains are characterized as having smooth surfaces, possible low scarps and intermediate albedos. Intervent plains are interpreted by Schaber (1982*b*) as being fallout deposits from volcanic plumes, interbedded with local flows and fumarolic materials.

Layered plains (Fig. 8.7) form smooth, flat surfaces and exhibit grabens and scarps up to 1700 m high, against suggesting compositions other than pure sulfur. The most conspicuous layered plains are found in the south polar region, although similar units occur throughout the mapped area. Transection by faults, overlap by various plains and possible erosional features suggest complex histories.

McCauley and co-workers (1979) note that mesas, hills

Figure 8.9 High-resolution image of Maasaw Patera showing large, complex caldera measuring 50 by 25 km, plus various flows which form a low-profile shield-like structure (Voyager 1 199J1+000).

◁ **Figure 8.8** View of a 100 km diameter caldera (A) near Creidne Patera and various flows. Bright zone around the base of one prominent flow (arrow) may be sulfur dioxide frost, as shown in Figure 8.7 (Voyager 1 75J1+000).

and remnants of subdued scarps beneath younger layers attest to multiple cycles of erosion and deposition. These disrupted, "etched" surfaces and irregular "fretted" margins cannot be attributed to fluvial or aeolian processes in the cold, tenuous atmosphere of Io, nor can ionic bombardment from Jupiter account for the features, given their relative youth. Rather, McCauley and his colleagues evoke a sapping mechanism in which liquid SO_2 is the dominant erosional agent. They suggest that a hydrostatic condition could be established if the crust were fractured, as might occur by faulting, driving molten SO_2 toward the surface. At the triple point for SO_2 part of the liquid would begin to crystallize and the system would expand, forming SO_2 vapor. Upon reaching the surface at or near the vent, the solid-fluid mixture would be released energetically at a velocity of ~ 350 ms^{-1}. At this velocity, the SO_2 "snow" could be sprayed as far as 70 km from the vent, although most of the material would fall closer to the vent. The numerous bright patches seen on Io (Fig. 8.8) are suggested to have formed by this mechanism. Thus, McCauley *et al.* suggest that newly formed scarps would be rapidly eroded by undercutting and slumping, leading to the formation of irregular margins, and that withdrawal of support from beneath the solid crust could

cause collapse, generating irregular surfaces. These processes would continue until the source of SO_2 was depleted locally.

Vent materials   Vent materials include all surface units that can be related directly to vents of various types. Vents include calderas (large, complex depressions; Fig. 8.9) and possible fissures. More than 300 vents have been identified, constituting about 5 percent of the total mapped surface area. The calderas average about 40 km across. Calderas on Io tend to be larger (up to 250 km across) and more frequent in equatorial regions than in the south polar region (maximum size is 100 km). They are much more randomly distributed on Io than on Earth, Moon on Mars, where the local and global tectonic framework seems critical to the location of volcanic vents. This suggests that the vents and related "hot spots" on Io are not controlled by strongly patterned convection cells.

Calderas show complex patterns of multiple eruptions and enlargements by wall collapse. The depth to the floor of the vents ranges from zero (level with surrounding surface) to more than 2 km. The relatively high, steep walls of some calderas may indicate that they are com-

Figure 8.10 Ra Patera, a large shield volcano located at 9°S, 126°W on Io. Measuring more than 600 km across, this is one of the largest volcanic constructs in the Solar System, nearly equal in size to Olympus Mons on Mars (Fig. 7.11).

Figure 8.11 Voyager image showing two unusual volcanic constructs. Apis Tholus (A) and Inachus Tholus (B). These disc-shaped volcanoes may have formed by very fluid lavas that spread symmetrically outward from their central vents and then solidified. Area shown is 600 by 800 km, centered at −17° latitude, 350° longitude (Voyager 1 71J1+000).

Figure 8.12 Image showing Kibero Patera located at 11°S, 300°W on Io. This complex volcano is composed of multiple flow units which originated from an irregular caldera in the lower left hand corner. Digitate flows extending to the right appear to represent "break-outs" of magmatic materials from the base of the complex flows. This feature measures 180 by 140 km (Voyager 1 105J1+000).

posed predominantly of silicate materials. Many of the floor units are very dark, suggesting the presence of molten sulfur. Imaging of some calderas by both Voyager 1 and Voyager 2 showed changes that occurred in the 4-month interval; part of Io's lava lake at Loki Patera may have crusted over, while new vents may have developed in other areas (Terrile *et al.* 1981).

Several different styles of flows have been identified on Io, including massive, coalescing units up to 700 km long (which may indicate high rates of effusion of relatively fluid materials) and narrow, sinuous flows up to 300 km in length which radiate from central vents. The flows may be composed of sulfur which melts in a low temperature (~ 115°C) and is very fluid. Furthermore, analyses of its thermal properties (Fink *et al.* 1983) suggest that, once mobilized, sulfur could flow long distances. Molten sulfur shows a variety of colors as it cools. Pieri *et al.* (1982; 1984) have analyzed Voyager images and, based on an assumption of sulfur composition, concluded that the observed colors of individual flows correlate with their distance from the vent.

Flows have accumulated to form several different types of volcanoes (Carr *et al.* 1979) including shields (Fig. 8.10), "discoid" volcanoes (Fig. 8.11), and more complex volcanoes that have multiple, digitate flows emanating from a more massive central construct (Fig. 8.12). Lack of topographic information, however, prohibits assessment of the exact morphology. Schaber (1982*b*) notes that shield-like volcanoes tend to be limited to the area between 45°S and 30°N, coinciding approximately with the equatorial band where active volcanic plumes are found (Strom and Schneider 1982).

8.2.2 *Volcanic plumes*

Nine active volcanic plumes (Fig. 8.4) were observed during the Voyager 1 encounter (Strom *et al.* 1979). As described by McEwen and Soderblom (1983), these plumes may involve more than one style of eruption. The most energetic sprayed sulfurous gases and solids as high as 300 km above the surface, at velocities up to 1 kms^{-1}. Pyroclastics rained down on the surface up to 600 km from the vent.

Of the nine active plumes observed during the Voyager 1 flyby, one cut off four months later when Voyager 2 imaged Io, and two new ones were observed. The volcanic plumes are driven by internal heating, possibly in a geyser-like mechanism involving SO_2 (Kieffer 1982). Johnson *et al.* (1979) estimate that upwards of 10^{10} tons of material are erupted from Io each year. Averaged over the entire area, the surface of Io is being buried at the rate of 100 m per million years. This figure closely correlates with the rates of surface renewal necessary to account for an absence of impact craters.

Combined with the various young flows observed on the surface, volcanic plumes indicate a highly active planetary object—one that has been "cooked" and differentiated through geologic time. Based on current activity, it is estimated that the entire mass of Io may have been recycled in its lifetime. Volatiles such as water and carbon dioxide have long been lost (Johnson and Soderblom 1982), while most heavier materials have sunk to form a core. Sulfur and various sulfur compounds, aided perhaps by bodies of silicate magmas, are constantly recycled, forming the complex surface observed today.

8.3 Europa

Of the Galilean satellites, Voyager imaging is poorest for Europa. The best resolution is only about 2 km per pixel, and the total coverage is rather small (Fig. 8.13). Despite these limitations, the surface of Europa displays features that pique the imagination.

Europa has an overall density of 3.0 g cm^{-3} and is considered to be differentiated into a silicate interior surrounded by a shell ~ 100 km thick which is composed of soft ice or water overlain by a veneer of ice. Depending upon the amount of heat generated by radioactive sources and by tidal stresses, like Io, Europa may have a partly liquid mantle. Its surface has the highest albedo of the Galilean satellites and is considered to be composed principally of water ice. However, the surface reflectivity and spectral characteristics are not uniform, suggesting that the ice has various impurities. These impurities could be derived from interior sources (such as silicates erupted onto the surface) or from exterior sources, such as sulfur emitted by Io and infall of meteoritic material.

8.3.1 *Physiography*

Surface features and terrain units have been mapped by Lucchita and Soderblom (1982) of the US Geological Survey, using Voyager images. In general, Europa displays little topographic relief, but has a variety of smooth and textured terrains of different colors, along with networks of grooves, ridges and other linear features. The part of the satellite imaged by Voyager is divided into two principal units, mottled terrains and various plains (Fig. 8.14). Mottled terrain is further divided into *brown mottled terrain*, which is moderately textured by small hills and forms sharp contacts with other units, suggesting that it is younger, and *gray mottled terrain*, which is smooth and has diffuse contacts. In general, the mottled terrain on the leading hemisphere is relatively bright, whereas the mottled terrain in the trailing hemisphere is darker. Lanzerotti *et al.* (1978) suggested that this

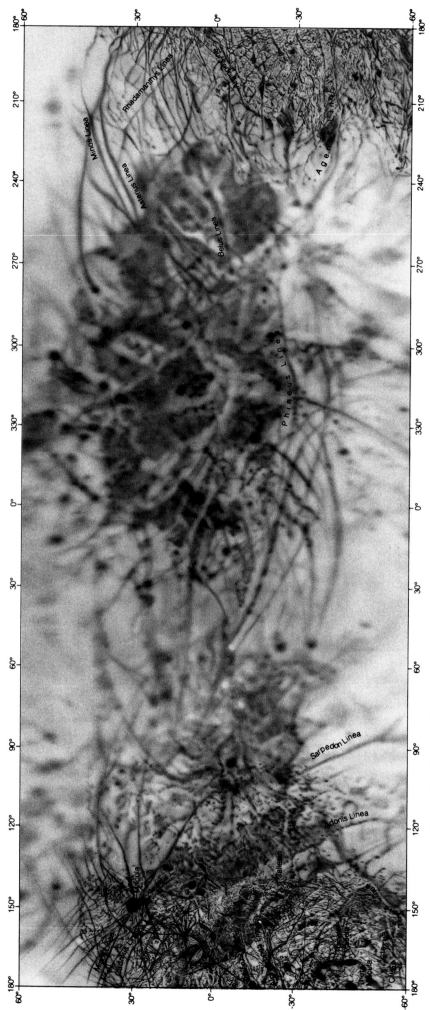

Figure 8.13 Shaded airbrush relief map of Europa showing prominent terrains and named features (courtesy US Geological Survey).

Figure 8.14 Voyager 2 image showing mottled terrain and various plains units (Lucchitta & Soderblom 1982): brown-mottled terrain (A), plains (B), gray-mottled terrain (C), and fractured plains (on the left hand side); also shown are wedge-shaped bands (D) and triple bands (E). Area shown is 2680 by 3160 km, centered at 10°S, 160°W (Voyager 2 1255J2-001).

difference results from the bombardment of ions originating from Jupiter's magnetosphere. Plasma within the orbital path of Europa travels faster than the satellite and would impact the trailing hemisphere, implanting ions in the ice and darkening the surface.

Plains units are subdivided into four types based on albedo and patterns of lineations. *Undifferentiated plains* are smooth surfaces cut by numerous linear features. This unit tends to be gradational with other units. *Bright plains* occur as polar deposits and have criss-crossing lineations that range from broad bands to faint streaks. Bright plains are the most highly reflective unit observed on Europa. *Dark plains* have a lower albedo than other plains units and it has been suggested that ice has been removed to leave a lag deposit of silicates. Alternatively, the low albedo could result from ions that are swept up by Europa, as described above. In either case, one would expect surfaces on Europa to darken with age. *Fractured plains* are found in the southwest part of the area mapped and are characterized by short, curved gray streaks and numerous brown spots.

In addition to the units described above, small spots and patches of brown and gray material occur on Europa. For example, Figure 8.15 shows Thrace Macula and Thera Mac-

ula, two brown patches in the southern hemisphere. Brown material is also found in small spots, along presumed fractures, and associated with some craters. Most investigators consider brown material to be of internal origin.

8.3.2 Craters

Lucchitta and Soderblom (1982) recognized two types of possible impact features on Europa. Class 1 craters are a few to a few tens of kilometers across and have raised rims, central peaks and ejecta deposits. Class 2 craters are large, flat, brown, circular areas 100 km or larger across. Craterforms of both types are very rare, at least in the areas photographed. Figure 8.16 shows various class 1 craters and includes a bowl-shaped crater, a central peak crater, a possible multi-ringed crater and a dark-halo crater. Figure 8.17 shows Tyre Macula, a class 2 feature. It consists of a circular area ~ 100 km across that has faint, concentric rings, but no apparent relief. Smith *et al.* (1989a) have described similar features on Ganymede as being impact craters that formed in ice-rich crusts and which have "relaxed" with time by viscous flow (see Section 8.4.2) or, alternatively, impacted into a material with a different rhe-

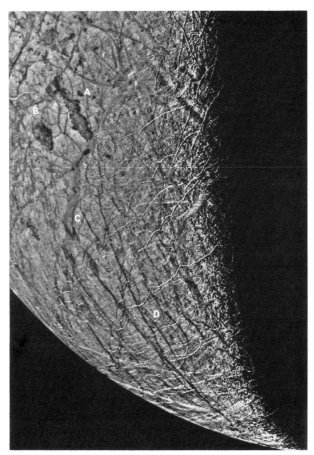

Figure 8.15 Image showing Thrace Macula (A), Thera Macula (B), gray bands (C) and cycloidal ridges (D). Thrace Macula is about 170 km long (Voyager 2 1372J2-001).

ology than silicate material, such as thin ice over a water substrate (Greeley *et al.* 1982).

In all, fewer than a dozen features that resemble impact structures have been found on Europa. The surface may be relatively young, with older craters having been destroyed or buried. Like Io, Europa experiences tidal stresses which may generate heat leading to volcanism. Furthermore, various lineaments may represent fractures which could have released liquids to the surface to flood and bury craters. However, calculations by Cassen *et al.* (1980) show that the tidal heating may be insufficient to melt internal ice, though interior water may still be liquid because it has never frozen. Instead, they suggest that the smooth and relatively uncratered surface may reflect the very earliest history in which liquid water was generated during initial differentiation and then persisted beyond the period of heavy bombardment.

8.3.3 Tectonic features

Global views of Europa (Fig. 8.18) display numerous linear features, some of which radiate from central spots.

These form patterns which bear a remarkable similarity to maps of Mars showing fanciful networks of "canali". Pieri (1980) classified several different linear and curved markings based on size, color and pattern. Along with various ridges, these were described by Lucchitta *et al.* (1981) and mapped by Lucchitta and Soderblom (1982), and include *dark wedge-shaped bands, triple bands* and *gray bands*.

Dark wedge-shaped bands occur south of the equator where they form a NW–SE trending zone. Individual bands can be 25 km across at their widest end and up to ~ 300 km long. They cut across other bands and hence are younger. Most investigators interpret these bands as representing a form of incipient plate tectonics. Helfenstein and Parmentier (1980) suggested that tidal stresses could rupture the surface, although Finnerty *et al.* (1980) consider the stresses to be inadequate for crustal failure. Ransford *et al.* (1980) offered an alternative model involving mantle upwelling and crustal extension.

Triple bands are among the most conspicuous features on Europa (Fig. 8.19). They consist of a pair of dark bands separated by a narrow, bright stripe or ridge. Triple bands range up to 18 km wide and > 1000 km long and trend northwest in the northern hemisphere and southwest south of the equator. Locally, some triple bands merge with single brown streaks. Others emanate from dark circular spots (due to impacts?). Most triple bands disappear where they enter brown mottled terrain. Helfenstein and Parmentier (1980) analyzed triple band patterns and concluded that they could have been generated by crustal stresses resulting from tidal deformation related to orbital eccentricities. Finnerty *et al.* (1980) proposed that the bands resulted from fracturing as a consequence of global expansion and that the fractures were filled with aqueous fluids and silicates to form breccia dikes. They suggest that relatively pure water was segregated and expanded upon freezing to form the central bright ridge. Gray bands (Fig. 8.15) are found in the south polar area. They are cut by all other lineations, including ridges, and are considered to be relatively old features.

Some Voyager images, taken along the terminator, enable subtle topography to be defined. A series of complex ridges was found, with the most extensive ridge occurring south of the equator. Some ridges are relatively straight, others (especially in the south polar area) are cycloidal (Fig. 8.15). From cross-cutting relations, the ridges appear to be the youngest features on Europa. Mapping by Dave Pieri (1980) shows that the ridges form small circles centered approximately on the anti-jovian point, which suggests that they may be related to tidal deformation.

8.3.4 Geologic history

Based on mapping and models of thermal evolution, Lucchitta and Soderblom (1982) have derived a tentative

a

b

c

d

Figure 8.16 Views of possible Class 1 impact craters on Europa (Lucchitta & Soderblom 1982). (a) Bowl-shaped crater 15 km in diameter (Voyager 2 1234J2-001), (b) 15 km crater with central peak and possible bright ejecta (Voyager 2 1368J2-001). (c) Multi-ringed crater (arrow), inner ring 30 km in diameter (Voyager 2 1219J2-001). (d) Dark-halo crater (arrow), 50 km in diameter (Voyager 2 1255J2-001).

Figure 8.17 Voyager 2 image of Tyre Macula (Class 2 craters); this 100 km diameter feature (arrow) seen in Figure 8.18 displays faint concentric rings and may be a form of impact crater (Voyager 2 1557J2-002).

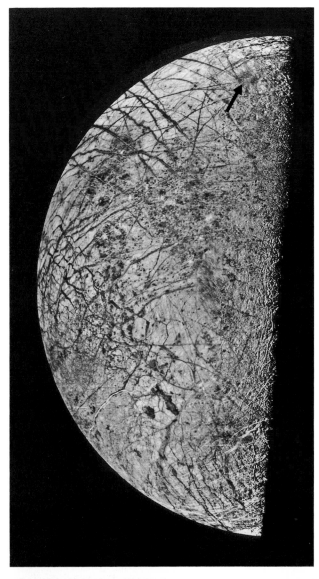

Figure 8.18 Mosaic showing various linear features which criss-cross Europa; note Tyre Macula's detailed structure (arrow).

Figure 8.19 View of "triple bands", ranging up to 18 km wide, which consist of two parallel dark zones separated by a central bright zone (arrow), which, in some cases, are ridges. Triple bands may be associated with global lineation patterns that form great circles on Europa (Voyager 2 20649.13)

geologic history for Europa that includes four general stages (from oldest to youngest):

1. Formation of bright plains material, probably by the freezing and solidification of an ice crust. Subsequent units appear to have formed at the expense of bright plains.
2. Formation of the dark plains, the fractured plains and possibly the gray mottled terrain. All involve the development of units which may result from intrusion of materials from subsurface sources or alteration of aging of other material.
3. Emplacement of brown material predominantly through fractures, but also localized as patches which could result either from eruptions or by impact excavation. Near the end of the emplacement, the brown mottled terrain was disrupted to produce complex topography.
4. Formation of ridges, possibly from dike-like intrusions.

Lucchitta and Soderblom emphasize that Europa displays a surface that is more the result of tectonism than of Earth-like "layercake" stratigraphy. Europa can be summarized (Morrison 1983) as a body having a rocky, lunar-like interior overlain by an icy crust that has been repeatedly fractured by tectonism and intrusion by water-rich materials.

8.4 Ganymede

Voyager views of Ganymede show a planetary crust that has been severely deformed and which may be the closest analog to Earth's plate tectonics. Voyagers 1 and 2 imaged approximately one-half of Ganymede's surface (Fig. 8.20) and returned photographs with resolutions as good as 500 m/pixel for some areas. Ganymede is larger than the planet Mercury and is the largest satellite in the Solar System. It has a density estimated to be slightly less than 2.0 g cm^{-3}, suggesting that it is about half water.

In addition to tectonic deformation, the surface of Ganymede displays terrain units of various albedos (Fig. 8.21). As discussed for Europa, lower albedo areas could result from continued infall of meteoritic material, or from the removal of ice to leave a lag deposit of dark materials. In any event, the greater impact crater frequencies observed on dark terrains indicate greater age.

8.4.1 Physiography

Preliminary mapping by Shoemaker *et al.* (1982) shows that Ganymede consists of two primary terrains (dark, cratered terrain and light, grooved terrain), two less extensive terrains (smooth plains and reticulate terrain) and basin terrains, plus various tectonic features and craters. The terrain units are subdivided primarily on the basis of albedo and crater frequency.

Dark, cratered terrain consists of low albedo, heavily cratered areas (Fig. 8.22) considered to be remnants of the ancient crust. This unit occurs as large, rounded areas or as small polygonal patches. The largest expanses of this unit include Galileo Regio on the leading hemisphere north of the equator, Nicholson Regio on the trailing hemisphere south of the equator, and Marius Regio which straddles the equator on the leading hemisphere.

Grooved terrain constitutes about 60 percent of the surface photographed. Most of both polar areas are composed of grooved terrain, at least of the areas imaged. Grooved terrain is characterized by subparallel grooves and ridges that are ≥ 100 km in length, separated by 3–10 km. Local relief from groove-to-ridge averages 300–400 m, but may be as great as 700 m. The grooves and ridges occurs as "bundles" a few tens of kilometers wide by hundreds of kilometers long. Bundles and sets of bundles display complex cross-cutting relations which can be used locally to derive sequences of formation.

In general, grooved terrain has formed at the expense of dark, cratered terrain, as shown in Figure 8.22, where bundles of grooves and ridges are seen to cut into darker, more heavily cratered terrain. Most investigators consider the processes of formation to involve rupture of ancient crust, perhaps accompanied by the release of liquid water or icy slush as a flood over the surface. Compression, extension, rotational shear or some combination of deformation have been proposed to explain the grooves.

Smooth plains occur as small patches and appear to be facies of grooved terrain (Shoemaker *et al.*, 1982). Smooth plains have about the same albedo as grooved terrain, but lack ridges and grooves (Fig. 8.23(a)). There is little topographic relief on the smooth plains, and they appear to be superimposed on adjacent units, suggesting that they may be relatively young flows.

Reticulate terrain appears to be a transitional unit between dark, cratered terrain and grooved terrain. The distinctive pattern (Fig. 8.23(a)) may result from two or more sets of grooves that are superimposed. In addition, the unit often includes numerous small hills which produce a hummocky texture.

Basins of probable impact origin also occur on Ganymede. Thus far, two such structures have been found, the largest of which is the Gilgamesh basin (Fig. 8.24). Although poorly defined, it has features suggestive of multi-ringed basins on terrestrial planets. Gilgamesh consists of a 150 km wide central depression filled with relatively smooth, featureless plains that are surrounded by rugged, mountainous ejecta deposits which decrease in relief radially outward. Secondary craters, many of

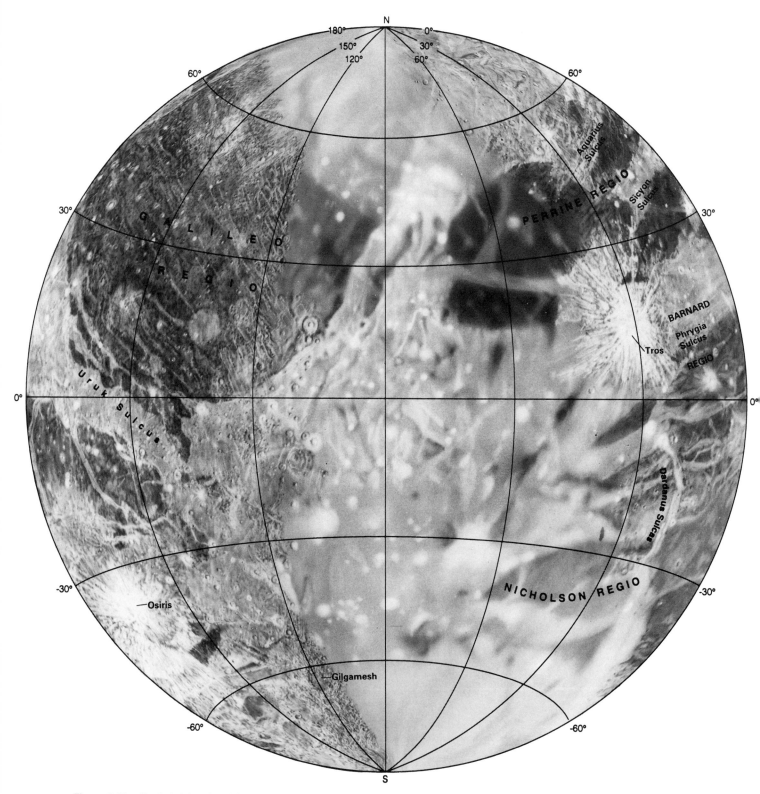

Figure 8.20   Shaded airbrush relief map of Ganymede showing various terrains and selected named features (courtesy of US Geological Survey).

Figure 8.21 Low-resolution view of Ganymede showing dark, cratered terrain, light grooved terrain, and bright halo craters. Circular dark zone in the upper right is Galileo Regio. Light, circular patches within Galileo Regio are palimpsests (Voyager 2 528J2-002).

Figure 8.22 View showing dark, cratered terrain in southern Galileo Regio and part of Uruk Sulcus (grooved terrain). Palimpsests (light, oval areas) are considered to be impact scars. One palimpsest (arrow) has been cut by a narrow band of grooved terrain. Also shown are furrows which cut the dark terrain. Note the dark material excavated by the crater on the lower left (A). Area shown is 1700 km across, centered at 9°N, 145°W (Voyager 2 264J2-001).

Figure 8.23a View showing two types of reticulate terrain near Galileo Regio. Reticulate terrain at (A) exhibits orthogonally intersecting grooves, while that at (B) is more irregular and hummocky. Various sets of grooved terrain crosscut dark terrain, older sets of grooved terrain, smooth plains (C), and the reticulate terrain. Arrow indicates a pit crater with a high-albedo central dome. Area shown is about 1500 km across, centered at 25°S, 176°W (Voyager 2 494J2-001).

Figure 8.23b Image of a domed palimpsest in the region of Nun Sulci, Ganymede. A swarm of secondary craters on the surrounding grooved terrain attests to the original presence of an impact structure near the center of the dome. The dome, more than 350 km across and ~ 2 km high at its center, may have formed by the combined processes of ice-volcanism and isostatic uplift. Image centered at 35°N, 328°W (Voyager 1 859J1+000).

which occur in chains, can be traced outward as far as 1000 km. A prominent ring occurs at a radius of 275 km; Shoemaker *et al.* (1982) consider this to represent the boundary of excavation for the basin. From stratigraphic relations and crater frequency distributions, Gilgamesh appears to have formed shortly after the development of the adjacent grooved terrain.

In addition to the Gilgamesh basin, an unnamed circular feature (Fig. 8.25) centered at ~ 7°S, 115° has been suggested by Shoemaker *et al.* (1982) as a possible impact feature. Superimposed on grooved terrain, its rim defines a diameter of about 200 km. Hummocky, ejecta-like deposits and secondary craters extend outward 150–450 km from the center. The floor of this structure is convex and has about the same radius of curvature as Ganymede, suggesting that hydrostatic equilibrium has been achieved.

8.4.2 Craters

One of the more striking discoveries of the Voyager mission is the unusual appearance of the impact craters on Ganymede and Callisto in comparison with craters on the terrestrial planets. With increasing diameter, the craters on the icy satellites become progressively flatter. A possible process responsible for this "flatness" was predicted

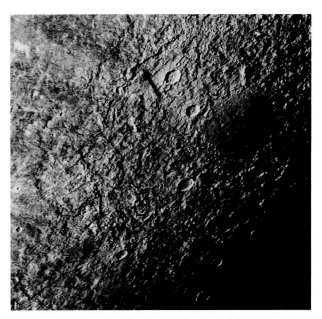

Figure 8.24 The Gilgamesh basin is defined by a smooth floor, 150 km across, in the center of the basin, surrounded by blocky terrain of 1.5–2 km relief. Prominent scarp (arrow) is considered by Shoemaker *et al.* (1982) to be the outer boundary of excavation. Area shown is centered at 61°S, 130°W (Voyager 2 527J2-001).

Figure 8.25 This unnamed western equatorial basin is one of the largest craterforms on Ganymede. It is 250 km across and highly "flattened" compared to the more recently formed Gilgamesh basin (see Fig. 8.24). Beyond the hummocky rim deposit to the upper left and lower left are swarms of secondary craters. Image centered at 4°S, 118°W (Voyager 2 552J2-001).

Figure 8.26 View of dark, cratered terrain showing small bowl-shaped crater (A) and central-pit craters (B). Area shown is in southeast Galileo Regio and is 600 km across (Voyager 2 546J2-001).

by Johnson and McGetchin (1973) well before the Voyager missions. They reasoned that plastic or viscous flow of icy crusts would reduce the relief of topographic features. Given the properties of ice and estimation of temperatures as a function of depth, Johnson, McGetchin and subsequent investigators (Paramentier and Head 1981) suggested that, following the impact excavation stage, crater floors first flattened, then bowed upward to a convex form as the crater rims subsided. As discussed by Passey and Shoemaker (1982), under these conditions, long wavelength features would be more affected by this **viscous relaxation** than short wavelength features. Thus, larger and deeper craters would deform faster and to a proportionately greater extent than smaller craters. For example, examination of craters on Ganymede shows that, while large primary craters are considerably "flattened", many of their secondary craters are bowl-shaped depressions.

Craters on Ganymede have been described by Passey and Shoemaker (1982) as including bowl-shaped, smooth-floored, central peak and central pit craters, plus an unusual craterform termed **palimpsests.** *Bowl-shaped craters* range in size from the limit in resolution to ~ 20 km in diameter (Fig. 8.26). Similar to small craters on terrestrial planets, they typically have depth-to-diameter ratios of 1:6 to 1:12. Many of the craters in this size range occur in chains and appear to be secondary craters. *Smooth-floored craters* have flat to slightly convex floors and range in diameter from ~ 20 to ~ 40 km. They appear to be transitional in morphology between smaller, bowl-shaped craters and large craters which have complex interiors. Floor convexity, or "up-bowing", seems to be more pronounced in craters formed in the ancient cratered terrains than elsewhere. *Central peak craters* are ~ 5 to ~ 35 km in diameter. Peaks have basal diameters of 1–5 km and are as high as 700 m. In contrast to craters on the Moon and Mercury, craters > 35 km on Ganymede lack central peaks. *Central pit craters* are commonly 16–120 km in diameter and essentially all craters on Ganymede larger than 40 km have pits. Generally, the diameter of the pit increases with crater size. Some pits have high albedo domes situated at their centers (Fig. 8.23(a)) which might be the result of ice diaperism (Malin 1980).

The term **palimpsest** refers to writing materials, such as parchment, that were reused in ancient times by removing the writing, although the original text was still faintly visible. In a similar vein, Smith and his fellow Voyager team members (1979*b*) applied the term to circular features they found on Ganymede that appeared to be the imprint of impacts (Figs. 8.22, 8.26 and 8.27). Palimpsests range in size from 50 to 400 km in diameter and occur as bright circular-to-oval patches with little topographic relief. The total area of palimpsests equals about 25 percent of the surface imaged. Most of them are zoned with an inner brighter area—thought to represent the excavation zone—and an outer rugged area representing part of the ejecta field. Some palimpsests form broad topographic domes (Fig. 8.23(b)) which might be the result of the combined processes of ice volcanism and isostatic upwelling (Squyres 1980).

Shoemaker and co-workers (1982) suggest that the high albedo interior zone of palimpsests represents relatively "clean" ice excavated from the subsurface. From the size relations and the inference of the excavation zone, they suggest that when most palimpsests formed, the lithosphere was about 10 km thick and was underlain by a water or slush mantle. They note that craters < 50 km in diameter do not have bright rims but exhibit bright floors, suggesting that the excavation bowl failed to penetrate through the lithosphere into the water mantle. Greeley *et al.* (1982) and Croft (1983) suggest that palimpsests and some shallow craters may be relatively pristine forms that result from impacts into ice which melts and deforms as part of the cratering processes. These craterforms are thought to represent the final stages of the period of heavy bombardment as the crust of Ganymede was solidifying.

Ejecta rays (both bright-ray and dark-ray) are prominent on Ganymede for some craters (Fig. 8.27). In general, rays are brightest when formed on grooved terrain and are darker for craters formed in dark, cratered terrain. As noted by Shoemaker *et al.* (1982), some rays exhibit abrupt changes in albedo along their length, some craters may have bright rays extending from one side and dark rays on the opposite side, while still other craters may have bright rays, but dark floors. Most of these variations can be attributed to differences in layers in materials being excavated.

Horner and Greeley (1982) found that many craters <90 km in diameter on Ganymede exhibit ejecta deposits that end abruptly in a scarp (Fig. 8.28), similar to some craters on Mars. We suggested that these could be a type of ejecta-flow crater in which ice-rich target material was melted during impact and at least partly emplaced as a fluidized mass.

Crater frequency distributions have been obtained for several areas on Ganymede (Greeley *et al.*, 1982; Passey and Shoemaker, 1982; Woronow *et al.*, 1982). Although the dark, cratered terrain (the oldest unit on the satellite) has about ten times the number of craters than lunar maria, it has fewer craters than the lunar highlands or the heavily cratered terrain of Mars. Crater counts on the grooved terrain are highly variable, with some south polar areas having about the same frequency as parts of the dark, cratered terrain.

8.4.3 Tectonism and volcanism

The prominent tectonic features on Ganymede include various grooves, fractures and grabens. Furrows and troughs cut the dark, cratered terrain, as shown in Figure

8.26. In Galileo Regio, these features can extend several hundred kilometers and are broadly arcuate, with the radius of curvature focused approximately on the anti-jovian point. McKinnon and Melosh (1980) have analyzed these features and suggested that they are regional grabens that formed in response to global stresses.

Grooves are typically flat-floored and occur in bundles, or packets, or subparallel scarps a few hundred kilometers long, although individual grooves may extend 1000 km. A few groove-sets exhibit some strike-slip displacement (Fig. 8.27), but for the most part, grooves appear to represent vertical displacement and are probably block faults.

Squyres (1980) and Parmentier *et al.* (1982) analyzed the global tectonic pattern exhibited on Ganymede and concluded that phase changes of ice could account for the fracturing and faulting. Ice occurs in several different crystal forms, called **polymorphs,** which have different densities and other physical properties as a function of temperature and pressure. These authors note that as Ganymede evolved, transitions from one ice phase to another would cause differences in global density and volume which could lead to a 7 percent increase in surface area. Expansion of this magnitude could have caused widespread normal faulting of the style exhibited in the grooved terrain. A subsequent analysis by Golombek (1982) shows that much less surface expansion could also produce the observed faulting in the grooved terrain. He estimates that the minimum thickness of the lithosphere was 5–9 km at the time of furrow formation and was about 4 km at the time of groove formation.

Crustal extension and normal faulting may also have released subsurface liquids. As discussed for Europa, this flooding can be considered a style of volcanism in which the "magma" consists of water and ice slush. The higher albedo of the grooved terrain and the presence of the smooth terrain support this interpretation.

8.4.4 Geologic history

Based on physiographic mapping, interpretation of surface-modifying processes and inferences of the interior properties and evolution, Shoemaker *et al.* (1982) have proposed the following history for Ganymede. The earliest record is characterized by the dark, cratered terrain which represents the early-formed lithosphere. During the early heavy bombardment (Shoemaker and Wolfe 1982), the lithosphere was too thin to preserve craters, but with time the crust thickened. Because of the physical properties of ice-rich materials, small craters would be preserved earlier than large ones. Palimpsests are thought to represent extreme cases of "relaxed" large craters or structures resulting from impacts into a plastic-like, deformable slurry underlying a thin, rigid crust and probably represent the final stages of the period of heavy bombard-

Figure 8.27 View showing impact crater with dark ejecta and bands of grooved terrain that have been cross-cut by younger grooved terrain (lower left corner); area shown is 2400 by 2400 km (Voyager 2 378J2-001).

Figure 8.28 High-resolution view showing craters which display ejecta deposits that terminate in a scarp similar to ejecta-flow craters on Mars. Area shown is 100 by 100 km, centered at 38°S, 125°W (Voyager 2 572J2-001).

ment. From preservation of rim structures, it is estimated that the lithosphere had thickened considerably by the time of formation of Gilgamesh, a multi-ring basin > 300 km in diameter.

Expansion of the core related to phase changes of the icy interior led to the development of grooved terrain through tectonic deformation and release of liquids to the surface. Formation of grooved terrain may also have resulted from upwelling convection cells within the mantle. As the lithosphere thickened, the formation of grooved terrain ceased. Local eruptions producing small patches of smooth terrain and impact cratering—principally by comets—mark the last major phases of Ganymede history. Surface modifications by infrequent impacts and ion bombardment continue today.

8.5 Callisto

Callisto is the darkest of the Galilean satellites, yet is still twice as bright as the Moon. Its density is only about 1.8 g cm^{-3}, making it the least dense of the Galilean satellites and suggesting that it has the greatest proportion of water. Voyager images show that the surface of Callisto is heavily cratered (Fig. 8.29) but has relatively little relief. Except for provinces related to large impact events, the surface viewed by the Voyagers is essentially uniform, consisting of relatively dark, heavily cratered terrain (Fig. 8.30).

Crater counts for Ganymede and Callisto (Fig. 8.31) show that Callisto is more heavily cratered than even the oldest terrain on Ganymede, despite the suggestion that the gravitational focusing effect of Jupiter should lead to more large craters on Ganymede. Thus, it is inferred that the surface of Callisto records a longer history than that of Ganymede.

The most prominent individual surface feature on Callisto is a multi-ring structure named Valhalla. As shown in Figure 8.32, Valhalla consists of a central bright zone about 600 km across surrounded by numerous concentric rings extending outward for nearly 2000 km from the center of the structure. The rings are spaced 20–100 km apart and ring spacing tends to increase with radial distance from the center. Mapping by Smith *et al.* (1989*b*) shows that there are relatively few superposed craters in the center of Valhalla and that the density of craters increases outward (Fig. 8.33(a)) until it reaches about the same frequency as the average of the whole satellite.

First recognized in a study by Hale *et al.* (1980), the rings of Valhalla may be divided into three zones based on morphology (Fig. 8.33(b)). The rings in the innermost zone consist of fairly continuous, narrow bands separated by wider bands of intermediate-albedo terrain. These bands are sinuous and scalloped in plain view and may be ridges, as are the innermost rings of the Asgard basin

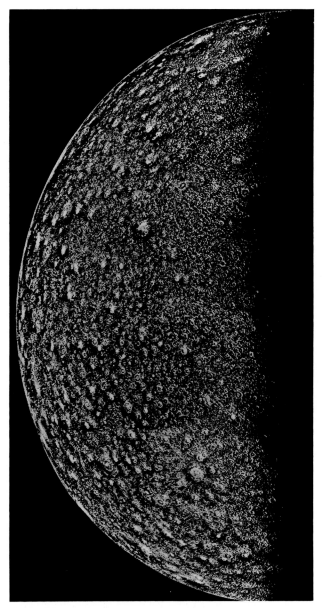

Figure 8.29 Mosaic showing the heavily cratered terrain of Callisto as viewed from a distance of 215,000 km (Voyager mosaic 260–586).

seen elsewhere on Callisto (Fig. 8.34). The width of this inner zone varies from ~ 200 km in the southeast to ~ 300 km in the northeast. The middle zone is characterized by weak to discontinuous ring development and varies in width from ~ 400 km in the southeast to ~ 200 km in the northeast. The outermost zone displays a wide variation in width and ring morphology. In the south and east the zone is > 500 km wide and consists of narrow, bright-floored sinuous troughs or furrows separated by wider, low-albedo terrain.

The resemblance of troughs or furrows to the furrows seen on the Galileo Regio of Ganymede (Figs. 8.22 and 8.26) led to the argument that the furrows on Galileo Regio were also formed by a large impact whose center

Figure 8.30 Shaded airbrush relief map of Callisto and selected place names (courtesy US Geological Survey).

Figure 8.31 Crater frequency distributions comparing various terrains on Ganymede with the substantially more heavily cratered surface of Callisto (from Smith *et al.* 1979b).

Figure 8.32 Voyager mosaic showing the Valhalla basin, defined by a central bright spot about 600 km in diameter surrounded by numerous concentric rings (Voyager mosaic P-21282).

has been destroyed by grooved terrain formation (McKinnon and Melosh 1980). To the northwest the outermost Valhalla ring zone increases greatly in width to 1400 km; the rings here consist of outward-facing scarps with dark, heavily cratered backslopes. Discrete light-toned bands which occurred below the scarps are thought to consist of material extruded from the scarp bases. In several places this light, extrusive material floods craters.

Jay Melosh (1982) developed a model which may explain the variation in ring morphology by determining the effect of the inward flow of the underlying asthenosphere toward the crater cavity as the initial impact cavity collapsed, producing a characteristic pattern of faults in the disrupted lithosphere as a function of distance from the point of impact (Fig. 8.35).

At least seven other multi-ring features have been recognized on Callisto (Passey and Shoemaker 1982). Asgard (Fig. 8.34), located at 25°N, 145°W has a central bright zone ~ 230 km across and is surrounded by concentric rings out to 800 km. Other, smaller features are also composed of central bright zones and concentric rings. The bright zones may be palimpsest-like features and the concentric ridges may have formed in the icy lithosphere as impact-generated adjustments.

Cassen *et al.* (1980), Passey and Shoemaker (1982) and others have considered the comparative evolution of Ganymede and Callisto. They note that in the early history both objects formed from materials leading to a water-silicate body. Heating from various sources, including impact cratering, radionuclides and tidal stresses led to differentiation and the formation of a "mud" core. In comparison to Ganymede, the smaller size of Callisto (and hence its lesser amount of radionuclides), the much smaller tidal effects and an inferred lower impact energy resulting from the reduced gravitational focusing of Jupiter because of the greater distance, all equate to heat being substantially lower on Callisto. Thus, its lithosphere probably thickened faster than Ganymede's, preserving the cratering record over a longer period and preventing tectonic deformation of the style which produced grooved terrain on Ganymede.

8.6 The Galileo mission

Many of the questions raised by the Voyager mission for the Jupiter system will be addressed by a mission

Figure 8.33a Highest-resolution images of the Valhalla basin, Callisto. The density of superposed craters increases outward from the center of the basin (at left). Mosaic is 3250 km across (Voyager 1 mosaic of 309J1+001 and 323J1+001).

INNER ZONE

TRANSITION ZONE

CENTRAL PALIMPSEST

flooded crater

OUTER ZONE

Figure 8.33b Sketch map of the morphologic zones into which the ring system of Valhalla basin was divided by Hale *et al.* (1980) and Melosh (1982).

scheduled for the mid-1990s. Named the Galileo mission, a dual spacecraft was launched from the shuttle in 1989 and will complete the journey of more than 10^9 km in six years. Prior to arrival, an entry probe will separate from the main spacecraft and will be sent on a trajectory to carry it into the atmosphere of Jupiter, measuring gas composition, pressure, temperature, wind velocities and other characteristics of the atmosphere. When the probe penetrates the upper atmosphere, it will travel $\sim$ 48 km s^{-1}, and it is estimated that frictional heat will ablate $\sim$ 100 kg of the protective shield as the probe descends. The probe will carry six instruments and is designed to function to a maximum pressure of about 30 bar, after which data return is extremely reduced.

The orbiter carries an imaging system, a near-infrared mapping spectrometer, and several other instruments. With a nominal lifetime of 20 months in orbit about the Jupiter system, the spacecraft will acquire about 2000 images of Jupiter and its satellites. During 10 orbits, numerous "encounters" or flybys will be made of the satellites. Although only one near flyby will be made of Io, images of $\sim$ 15 m resolution will be obtained for a small part of the planet. More distant passes will provide $\sim$ 1 km resolution images that will enable monitoring of volcanoes. Several flybys will be made of Europa. Galileo coverage for this satellite will be especially interesting because Voyager images are relatively low resolution, yet show many intriguing landforms. Ganymede and Callisto will also be extensively imaged with some regions observed at $<$ 100 m resolution.

In general, Galileo should provide images with about an order of magnitude higher resolution than Voyager. Some encounters, such as the Io flyby, will yield even higher-resolution pictures. In addition, Galileo will image areas not seen by Voyager, expanding our knowledge of the global characteristics of all the Galilean satellites.

Figure 8.34 Voyager image of Asgard basin, Callisto. This multi-ring system is very similar to Valhalla but is only about 1500 km in diameter. Low sun angle reveals that the inner rings are ridges which appear to be flat-topped. Image is about 900 km across, centered at 14°N, 126°W (Voyager 1 526J1+001).

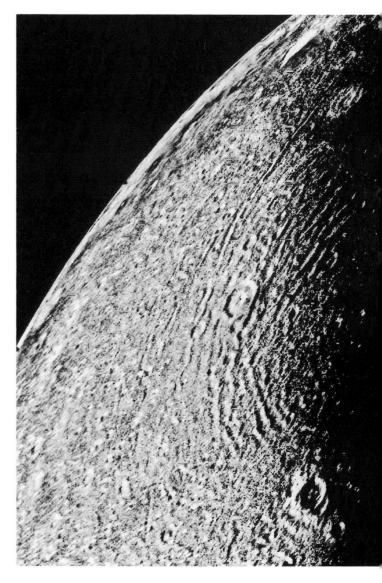

Figure 8.35 Lithosphere-asthenosphere response to the impact which formed the Valhalla basin, Callisto. Note that flow of the asthenosphere, associated with the infilling of the transient cavity, placed the lower lithosphere in motion having different directions of stress with radial distance from the center of impact. This results in different physiographic zones that are concentrically arranged around Valhalla (from Melosh 1982).

transitional zone

inner zone composed of ridges

site of transient cavity

outer zone composed of narrow furrows

south

outer zone composed of outward-facing scarps

north

rigid lithosphere

SOLID MANTLE

FLUID-LIKE ASTHENOSPHERE

9 The Saturn system

9.1 Introduction

Prior to the Voyager flybys in 1980 and 1981 (Table 8.1), the satellites of Saturn appeared as little more than tiny spots of light, even when seen through the most powerful Earth-based telescopes. The Voyager mission not only transformed these spots into objects available for geologic study, but also enabled discovery of at least eight additional satellites, bringing the total known number to 17 (Fig. 9.1). These represent a remarkable diversity of objects, including various icy moons (many of which show signs of repeated resurfacing), the only known satellite in the Solar System to have an appreciable atmosphere (Titan), and a host of "small bodies" (a few tens of kilometers in size) which can be compared with the moons of Mars (Phobos and Deimos) and the innermost moon of Jupiter (Amalthea).

Although Pioneer 11 preceded the first Voyager to Saturn by a little over a year, most of the instruments carried by the Pioneers were designed to take measurements, such as the presence and strength of magnetic fields. However, a simple imaging system also enabled the discovery of a new satellite, as well as a new ring of Saturn—the F Ring—which lies a few thousand kilometers beyond the A Ring. Pioneer also discovered a second satellite, but not by imaging: the spacecraft nearly collided with an object estimated at the time to be ~ 200 km in diameter. Analysis of data related to magnetic fields showed that the spacecraft passed through the magnetic "wake" of the unknown satellite which was imaged later by Voyager.

Nearly all the geologic data on the saturnian satellites come from the Voyager missions. The paths of both Voyagers were planned to use the gravity of Jupiter to "sling-shot" them on their way to Saturn. Voyager 1 flew closer to Jupiter and received a greater boost in speed that allowed it to arrive at Saturn in November 1980, nine months earlier than Voyager 2. Aimed to make a close pass by Titan, Voyager 1 also viewed the dark (shaded) side of the rings and obtained high-resolution views of Mimas, Dione and Rhea. However, programming the flight path for a close flyby of Titan meant that the trajectory would carry Voyager 1 out of the ecliptic plane of the Solar System and prohibit it from journeying onward to the other outer planets.

The path of Voyager 2 was planned so that it encountered Uranus in 1986 and Neptune in 1989. Although this requirement placed some constraints on its passage through the Saturn system, the trajectory nicely complemented that of Voyager 1. Voyager 2 made a pass of the illuminated side of the rings and provided close-up views of the satellites Iapetus, Hyperion, Enceladus and Tethys. Unfortunately, during the close approach to Enceladus and Tethys, the scan platform which carries the camera and other instruments became stuck and was unable to move in an azimuth (back and forth) direction. Consequently, the highest-resolution images of these satellites, as well as stereoscopic views of the F Ring were lost because the cameras were pointing in the wrong direction. At first blamed on the possible impact of particles associated with the ring system, analysis showed that the jamming of the scan platform resulted from leakage of lubricant. Subsequent study by spaceflight engineers shows that the platform could be moved, if done very slowly, and that imaging was successfully accomplished when Voyager 2 arrived at Uranus and Neptune.

Despite the disappointment caused by loss of the highest-resolution images, the Voyagers yielded a wealth of information on Saturn, its rings and the satellites, as described by Smith et al. (1981; 1982) and reviewed by Morrison (1982b) and Soderblom and Johnson (1983). Review papers on Saturn are given in Gehrels and Matthews (1984). Reviews on the rings are provided in Greenberg and Brahic (1984), and on the satellites in Burns and Matthews (1986).

Like Jupiter in many respects Saturn also is a miniature Solar System, but with some important differences. For example, in both the Jupiter system and the Solar System in general, the densities of the orbiting bodies decrease with radial distance from the primary object. Analysis of the densities of saturnian satellites shows no such relationship (Table 2.1); rather, water-rich and rocky objects seem to be randomly distributed throughout the system.

Estimates of the satellite densities indicate that they are mixtures of rock and water ice (up to 70 percent ice by mass). Titan, a Mercury-sized object, is the most dense at 1.9 g cm^{-3} and is thought to be about half rock and half ice. Analysis of reflectance spectra shows that the surfaces of all the satellites are dominated by water frost. In fact, Enceladus is the brightest object in the Solar System, being over ten times more reflective than Earth's Moon.

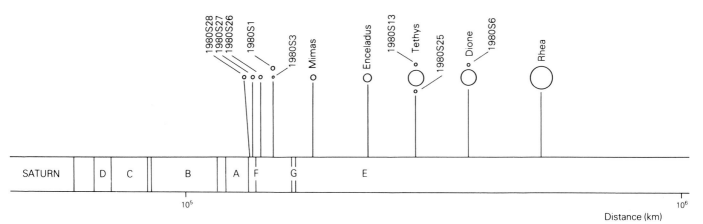

Figure 9.1 Diagram showing the known satellites of Saturn drawn to their correct relative sizes and showing their positions within the saturnian system. The distance scale is logarithmic, and the distances are from the center of Saturn. Also shown are the locations of the major components of the ring system.

In this chapter each of the major satellites is briefly described and illustrated. A section then outlines the processes that appear to have modified their surfaces.

9.2 Geomorphology of the satellites

The nine largest satellites were discovered telescopically, some as early as 1655. Most of their names were proposed by Sir John Herschel in the 1700s. In Greek mythology, Dione, Rhea, Tethys, Mimas, Enceladus, Titan and Phoebe were giants, while Hyperion and Iapetus were brothers of Saturn.

All of the satellites but two (Hyperion and Phoebe) are in synchronous rotation, i.e. they keep the same side facing toward the primary, as do Earth's Moon and the Galilean satellites. All but two travel in circular prograde orbits in the equatorial plane, which is also the plane of Saturn's rings (Fig. 9.2). The exceptions, Iapetus and Phoebe, travel in paths inclined to the equatorial plane, and Phoebe travels in a retrograde direction. The innermost satellites have orbits interspaced between some of the prominent rings (Fig. 9.1).

Image resolution ranges from very poor to a couple of kilometers for most of Saturn's major satellites. Thus, terrain mapping is difficult for many of the satellites and caution must be exercised in comparing geologic processes and histories because of the uneven image quality for the different objects.

9.2.1 Mimas

Discovered in 1789 by Herschel, Mimas is a ~ 390 km size object lying between the G Ring and the E Ring of Saturn. During the flyby of Voyager 1 it was imaged at a resolution of about 1 km and shows a cratered surface (Fig. 9.3). Although the distribution of craters is not uniform, most of the surface of this icy object appears

to be in equilibrium for craters < 30 km; that is, craters up to this size are destroyed at the same rate as they are formed. Most of the craters are deep, bowl-shaped depressions (Fig. 9.4), and many larger than about 20 km have central peaks. Well-defined ejecta deposits and bright rays are apparently lacking, although the high albedo of the surface would make rays difficult to see, even if they do exist.

The most striking aspect of Mimas is an impact crater, named Herschel, found on the leading hemisphere (Fig. 9.5). With a diameter of 130 km, it is about one-third the diameter of Mimas. This crater is nearly 10 km deep, with a central peak that rises some 6 km from the crater floor. Larger than Tycho and Copernicus on the Moon (Fig. 4.31), Herschel must be near the maximum size impact that the satellite could survive without shattering apart.

In addition to craters, Mimas displays grooves up to 90 km long, 10 km wide, and 1–2 km deep (Fig. 9.4). The small size of Mimas may preclude internal differentiation (supposedly there would be insufficient radionuclides to melt the ice) and related tectonic deformation, and these grooves could be related to the formation of the crater Herschel or some other impact event. The trailing hemisphere of Mimas shows local clusters of hills 5–10 km across and ~ 1 km high which may be ejecta deposits.

9.2.2 Enceladus

Like Mimas, Enceladus was discovered by Herschel in 1789. Voyager 1 images were of relatively low resolution and showed this small moon to be a bright, smooth sphere. Voyager 2 obtained images of ~ 1 km resolution and revealed complex terrains (Figs. 9.6 and 9.7) which, in many respects, are similar to those of the Galilean moon, Ganymede.

Given the similar sizes of Mimas and Enceladus and their positions relative to Saturn, before the Voyager mission it was generally thought that the surfaces of these

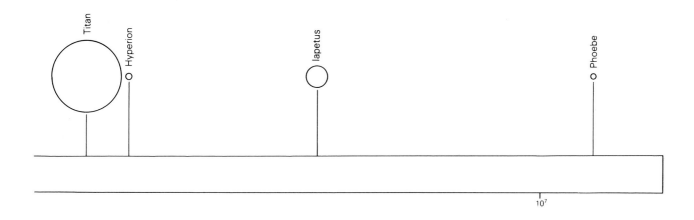

Titan

Hyperion

Iapetus

Phoebe

10^7

Figure 9.2 Voyager image showing prominent dark "spokes" in the outer half of Saturn's B ring (Voyager 2 JPL P-23881).

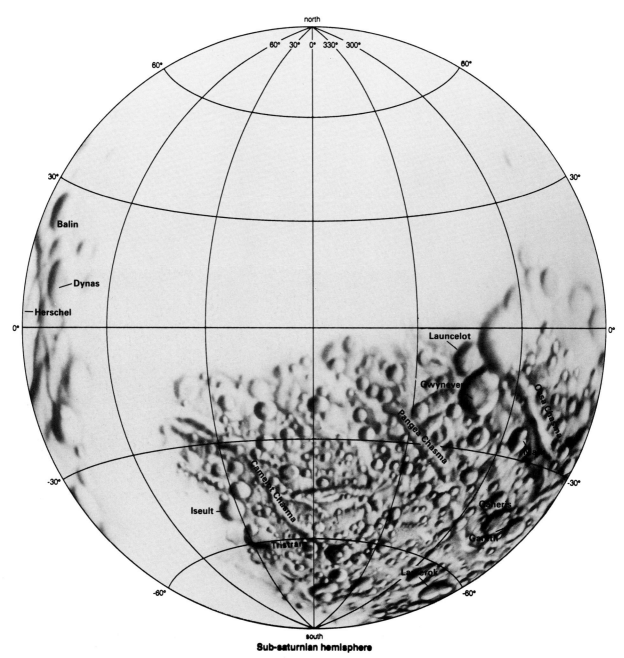

Figure 9.3 Shaded airbrush relief map of Mimas (courtesy US Geological Survey).

Anti-saturnian hemisphere

223

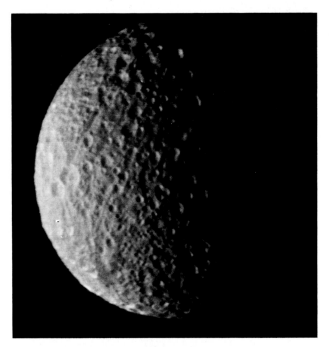

Figure 9.4 High-resolution image of Mimas showing the south polar region and prominent grooves (Voyager 1 25S1 + 000).

Figure 9.5 Voyager image of Mimas showing impact crater, Herschel (Voyager 1 JPL P-23210).

two satellites would be similar. Yet, they represent two extremes. While Mimas preserves a cratering record back to the time of its formation and shows no sign of resurfacing, Enceladus displays a complex geologic history.

Preliminary terrain mapping by Smith *et al.* (1982) and Passey (1983) shows various smooth plains, ridged plains and cratered terrains (Fig. 9.9). *Cratered terrain* is characterized by abundant craters 10–20 km in diameter and is subdivided into two units—in one, the large craters are shallow (perhaps as a result of viscous deformation), while the second is younger terrain in which the craters are not flattened and which may reflect a change in the properties of the lithosphere to a more brittle material. Passey (1983) has carefully analyzed crater profiles to infer properties of the lithosphere of Enceladus and to model the evolution of the interior. He concluded that Enceladus's lithosphere must have a mixture of ammonia ice and water ice, in contrast to the relatively pure water ice of the Galilean satellites.

Cratered plains, centered at 30°N, 345°W, are characterized by bowl-shaped craters 5–10 km in diameter; *smooth plains* are lightly cratered and display a rectilinear pattern of grooves in some areas; *ridged plains* are characterized by complex, subparallel ridges up to 1 km high with interspersed smooth plains. Much of the ridged plains terrain occurs on the trailing hemisphere and appears to have formed at the expense of older cratered terrain.

The landforms seen on Enceladus exhibit tectonic and resurfacing processes unexpected for such a small object. The various episodes of resurfacing probably involved extrusion of water or water ice slurries. The grooves are considered to be grabens resulting from extension and brittle failure of the lithosphere. The ridges, which are similar to those on Ganymede, could result either from compression, from upwelling by convection from the interior, or from expansion of freezing water which intruded fractures.

What could have generated the tectonic deformation and heat to release water? Steve Squyres and colleagues (1983) have analyzed the problem, taking into account the potential sources of heating. Because of Enceladus's small size, neither primordial heat generated by accretionary impact nor radionuclide heating could account for its activity—Enceladus would have frozen solid very early in its history. It is unlikely that there are sufficient radioactive elements unless the satellite's composition is extremely unusual in comparison to known Solar System elemental abundances. The most likely source is tidal heating, similar to that which affects Io. Enceladus has an orbital eccentricity forced by a resonance with Dione. This eccentricity would "push–pull" the satellite and would, in turn, generate tidal heating. Although Squyres *et al.* (1983) conclude that greater eccentricities than present values are required to melt an initially frozen body, especially if pure water ice is involved, they note that the inclusion of hydrates of ammonia or a methane clathrate could lower the melting temperature substantially (by 100°C or more).

It has even been suggested that Enceladus may experience active volcanism. Although there is no direct evidence of eruptions, it has been noted that the densest

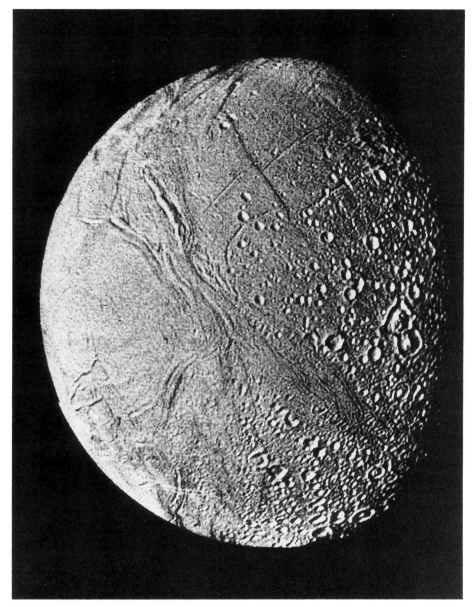

Figure 9.6 Image of Enceladus showing cratered terrain, grooved terrain, and smooth terrain; note the lateral offset along a long fracture. The smooth terrain at the subsolar point (left side) may be an illusion due to the effects of illumination (Voyager 2 JPL P-23956).

part of Saturn's E Ring is coincident with the orbit of Enceladus (Fig. 9.1), suggesting that the satellite could be the source of the material which makes up the ring (Smith *et al*. 1982).

9.2.3 Tethys

Tethys is more than twice the diameter of Enceladus and displays landforms indicative of tectonic deformation and resurfacing, although these features are not as extensive as on Enceladus. The two most striking surface features of Tethys are Odysseus, an impact crater 400 km in diameter (about 40 percent of the diameter of the satellite), and an enormous canyonland, Ithaca Chasma, which

wraps three-quarters of the way around the globe (Fig. 9.8). The general geology of Tethys has been described by Moore and Ahern (1983) who mapped the major terrains (Fig. 9.10). *Hilly cratered terrain* is the oldest recognized unit and is characterized by rugged topography and a high frequency of craters > 20 km in diameter, most of which are degraded. *Plains terrain* is centered at 10°S latitude, 310° longitude, coincident with the trailing hemisphere. It consists of less rugged, more sparsely cratered surfaces in comparison with the hilly cratered terrain. In addition, plains terrain displays faint lineations and variations in its albedo.

The crater Odysseus is equal in size to Mimas and occurs on the leading hemisphere. It is a relatively shallow feature with a ring-like central peak complex and a floor

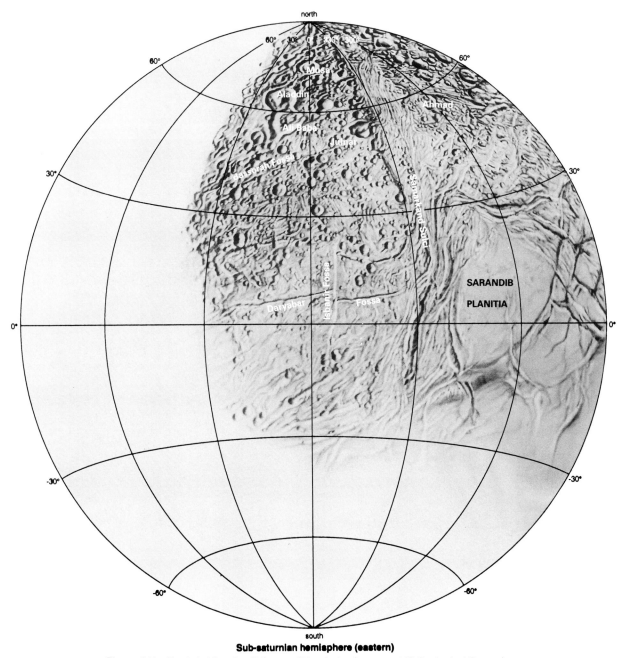

Figure 9.7 Shaded airbrush relief map of Enceladus (courtesy US Geological Survey).

226

Anti-saturnian hemisphere (western)

Figure 9.8 Shaded airbrush relief map of Tethys (courtesy of US Geological Survey).

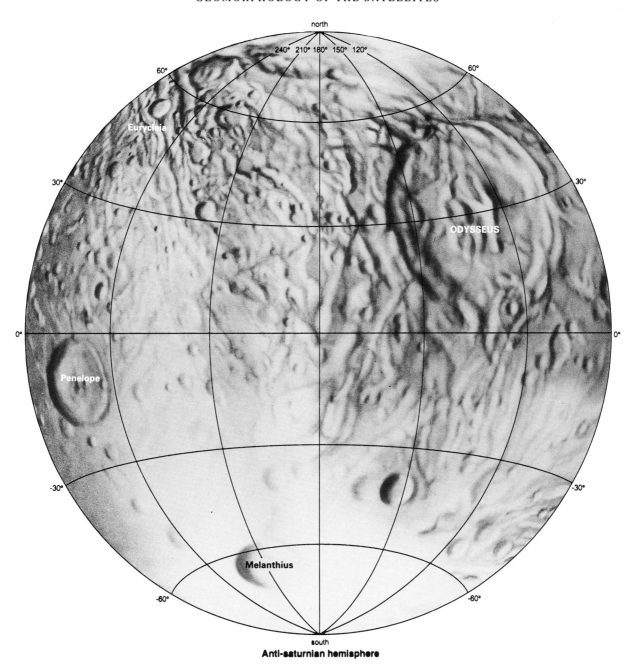

north
240° 210° 180° 150° 120°
60° 60°
30° 30°
Eurycleia
ODYSSEUS
0° 0°
Penelope
-30° -30°
-60° -60°
Melanthius
south
Anti-saturnian hemisphere

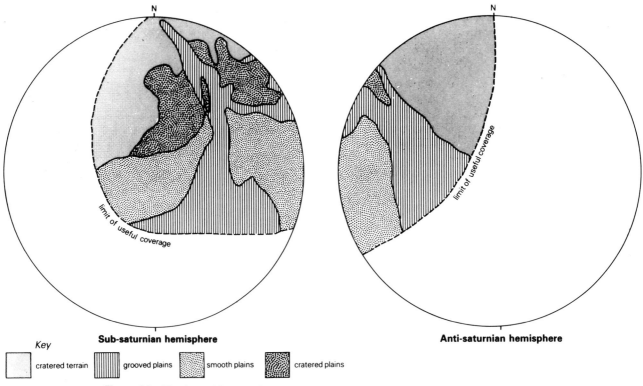

Figure 9.9 Physiographic map of Enceladus (from Passey 1983, copyright © Academic Press).

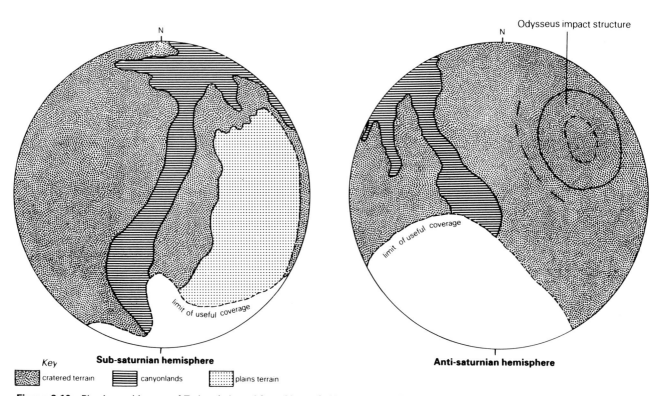

Figure 9.10 Physiographic map of Tethys (adapted from Moore & Ahern 1983, *J. Geophys. Res.,* copyright American Geophysical Union).

Figure 9.11 Views of Tethys showing impact crater Odysseus. The floor of Odysseus matches the curvature of the satellite (Voyager 2).

that matches the planetary curvature (Fig. 9.11). Smith *et al.* (1982) note that the low relief of Odysseus may be a consequence of viscous deformation of an icy lithosphere. Soderblom and Johnson (1983) propose that the impact occurred when the interior of Tethys was partly liquid or still "soft" (plastic-like) ice. Moore and Ahern (1983) note that only about 1–2 percent of the ejecta from Odysseus would have escaped the satellite to be placed in orbit about Saturn, and that much of the material would be swept up later by Tethys.

An impact of the size to produce Odysseus would have had a substantial effect on such a small satellite. Seismically induced degradation in the antipodal zone of the moon may have generated terrain similar to the hilly and lineated terrain antipodal to the Caloris Basin on Mercury. However, Moore and Ahern note that the smooth plains—which evidently post-date Odysseus—occur in the area antipodal to Odysseus and would have buried the seismically jostled terrain, even if it had formed.

Ithaca Chasma is a branching, terraced canyon system more than 1000 km long, 100 km wide and up to 4 km deep (Figs. 9.12 and 9.13). Parts of the canyon rim are raised 500 m above the surrounding terrain. As mapped by Moore and Ahern (1983), many lineaments splay off Ithaca Chasma, forming V-shaped patterns pointing northward. The canyonland may have been created as some sort of lithospheric response to the impact that formed Odysseus. It has been observed by Smith *et al.* (1982) and Moore and Ahern (1982) that the canyonland comprising the Ithaca Chasma complex generally lies on a great circle which is perpendicular to a radius-line from the center of Odysseus (Fig. 9.14). Moore and Ahern (1983) invoked seismic disruption from the Odysseus impact as the rifting mechanism that formed the canyonland. They noted that if the Odysseus impact caused

Tethys to become temporarily oblate, this could induce near-surface tensional fracturing along a narrow region that is presently occupied by the canyonland. Alternatively, models of the thermal evolution of Tethys which begin with freezing of water shows that the volume would expand $\sim$ 10 percent, which equals an increase in surface area of $\sim$ 7 percent. Tectonics of this magnitude could easily account for the extension to form Ithaca Chasma but this model fails to explain why the canyonland occurs only in a narrow zone.

9.2.4 Dione

Dione, discovered in 1684 by Cassini, is about the same size as Tethys but is considerably more dense. At 1.43 g cm^{-3}, it is second only to Titan in density, indicating a higher proportion of rocky material in comparison to the other icy satellites. Dione also shows a wide range of albedo patterns, including bright, wispy markings (Fig. 9.18). Geologic mapping by Plescia (1983) and Moore (1984) shows that Dione can be classified into cratered terrain and various plains units (Fig. 9.15). *Heavily cratered terrain,* consisting of rough surfaces having numerous craters > 20 km in diameter, has the greatest areal extent of the region imaged by Voyager. *Cratered plains* have a lower frequency of large craters in comparison to heavily cratered terrain, whereas *smooth plains* have very few craters or other topographic features.

The trailing hemisphere of Dione is distinguished by a network of intensely bright streaks set on a dark background. These wispy features are composed of, or pass into, narrow bright lines (Fig. 9.16). In addition an elliptical $\sim$ 240 km feature, named Amata, occurs in the center of the complex and may be an impact scar. The bright wispy marks, however, do not resemble bright

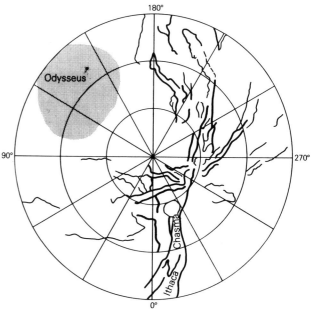

Figure 9.12 Voyager view of Tethys showing Ithaca Chasma (arrow), a plains unit (lower right) and heavily cratered terrain (Voyager 2 JPL P-24065).

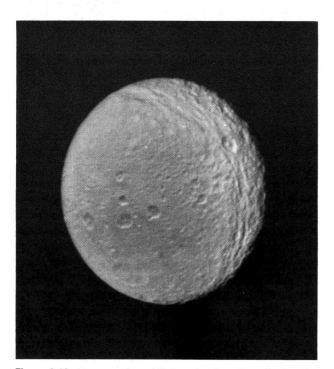

Figure 9.13 Voyager view of Tethys showing a low albedo strip running north-south across the plains unit in the center of the image. Ithaca Chasma is visible near the terminator (Voyager 2 JPL P-23948).

Figure 9.14 Diagram of north polar region of Tethys showing the relation of Odysseus to various lineaments (from Moore & Ahern 1983, *J. Geophys. Res.,* copyright American Geophysical Union).

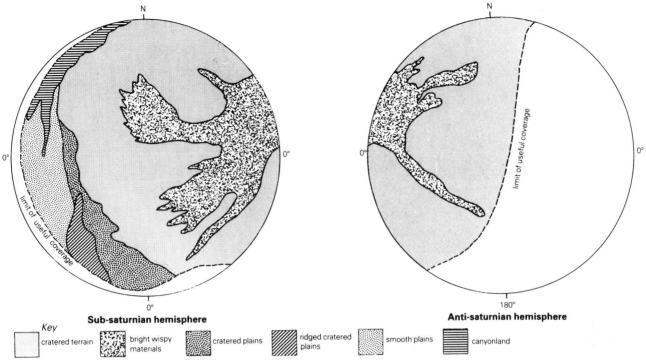

Sub-saturnian hemisphere

Anti-saturnian hemisphere

Key

cratered terrain	bright wispy materials	cratered plains	ridged cratered plains	smooth plains	canyonland

Figure 9.15 Physiographic map of Dione (modified from Plescia 1983, copyright © 1983 Academic Press).

Figure 9.16 Voyager image of the trailing hemisphere of Dione showing bright, wispy streaks crossing the trailing, low-albedo hemisphere (Voyager 1 1182S1-001).

Figure 9.17 Voyager image of the trailing hemisphere of Dione showing heavily cratered terrain, the apparent association of bright wisps with troughs, and the parallel groups of some of the bright wisps' related lineaments (Voyager 1 62S1 + 000).

Figure 9.18 Shaded airbrush relief map of Dione (courtesy US Geological Survey).

north

south

Anti-saturnian hemisphere

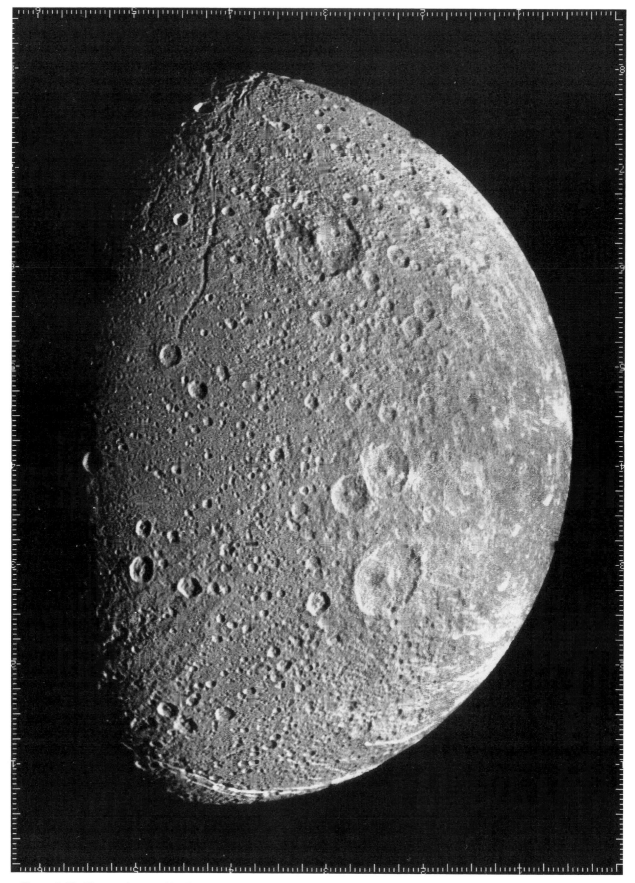

Figure 9.19 Voyager image of leading hemisphere of Dione showing troughs and trough networks in the northern hemisphere.

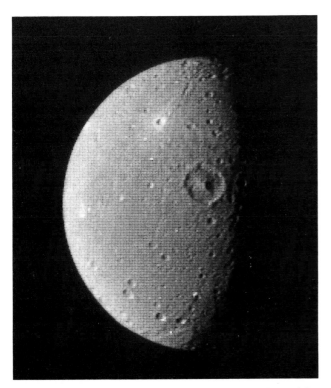

Figure 9.20 Voyager 2 image of Dione showing area beyond the terminator of Figure 9.19. Note the westward extension of the plains and the only unequivocally identifiable ray crater (Creusa) seen on a saturnian satellite (upper portion of disk) (Voyager 2 1239S2-001).

rays from impact craters but, rather, follow irregular paths. Smith *et al.* (1981) suggest that these markings could represent deposits of frost-like material formed by the explosive release of volatiles from the interior through fractures.

The largest craters on Dione occur on the trailing hemisphere (Fig. 9.17) and are as large as 200 km across. Craters > 100 km in diameter are common in the heavily cratered terrain. Most of the large craters have terraced walls and central peaks (Figs. 9.19 and 9.20), while many of the smaller (< 15 km) craters are simple, bowl-shaped depressions. As is the case for all the saturnian satellites, clearly defined ejecta deposits (other than crater rays) are lacking, although this could be the result of poor image resolution. In general, craters are more shallow on Dione than on Tethys.

Tectonic features on Dione include ridges, scarps, troughs and some crater chains, as described by Jeff Moore (1984). Ridges are 50–100 km long and <0.5 km high. These features were mapped by Plescia (1983) as lobate deposits, which he suggested to be similar to martian ejecta flow deposits, although a source impact crater was not identified on Dione. Scarps up to 100 km long occur in the heavily cratered terrain. These may be the result of faulting or they could be landforms associated with mass wasting. Troughs range in length from 30–

100 km, but some exceed 500 km, and most have a pit crater at one or both ends (Fig. 9.19). Troughs may form networks, as in the smooth plains, or complex branching systems. Chains of craters, found principally on the plains, are as long as ~ 100 km and 10 km wide and may be secondary impact craters, volcanic craters or aligned pit craters of internal origin, such as collapse.

Moore (1984) analyzed the landforms on Dione and proposed the following evolutionary sequence. After the formation of a brittle lithosphere following accretion, Dione experienced global expansion resulting from heat generated by radionuclides and tidal stresses. This expansion produced a pattern of global lineaments prior to the end of heavy meteoritic bombardment. Subsequently, an ammonia–water melt was produced in the interior and was erupted onto the surface to form plains. Cooling of the interior and/or a phase change led to compression of the surface, resulting in thrust or high-angle reverse faults which formed ridges. Later events involved light impact cratering, primarily by cometary objects.

9.2.5 Rhea

Voyager 1 passed within 59,000 km of Rhea and returned the highest resolution image available from any of the saturnian satellites, about 500 m/pixel for part of the north polar region (Fig. 9.21). Except for Titan, Rhea is the largest of the satellites, having a diameter of 1530 km. Like Dione, its trailing hemisphere is dark and has bright wispy markings (Fig. 9.22), whereas the leading hemisphere is uniformly bright, except for one large, diffuse, very bright feature thought by Smith *et al.* (1981) to be ejecta deposits. Cratered terrains are dominated by large, degraded craters and resemble the lunar highlands. None of the craters show the "flattening" to the degree of those on Ganymede. Most of the craters ⩾ 25 km have central peaks and some craters show very bright patches on the walls, which Smith *et al.* (1981) attribute to fresh ice deposits or other fresh materials exposed by slumping (Fig. 9.23).

The general surface features on Rhea (Figs. 9.24 and 9.25) have been described by Moore and Horner (1984) who note the presence of a ~ 450 km multi-ring basin (Fig. 9.26). The outer ring appears to consist of two concentric, inward-facing scarps, with the inner scarp being much more prominent and the outer scarp being discontinuous in places, similar to Odysseus on Tethys. The spacing of the scarps is ~ 10–15 percent of the diameter of the basin. The inner ring is slightly more than half-way between the center of the basin and the inner scarp and is composed of ridges or hills rather than scarps. The outer rim scarps may be analogous to mega-terraces, as proposed for some multi-ringed basins on the terrestrial planets.

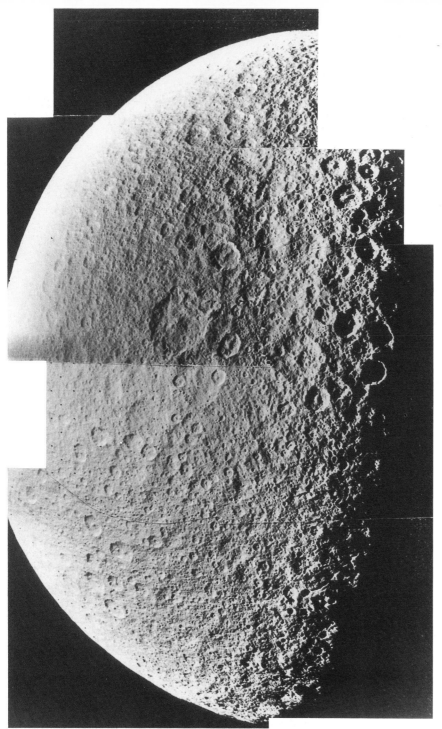

Figure 9.21 Mosaic of Voyager images for north polar region of Rhea showing both ancient cratered terrain (on which occur coalescing pits and linear troughs) and resurfaced terrain dominated by smaller, population II craters (and on which scarps and ridges can be seen) (Voyager 1 JPL P-23177).

Figure 9.22 View of Rhea showing differences between the leading (bright) hemisphere and trailing (dark) hemisphere (Voyager 1 1558S1-001).

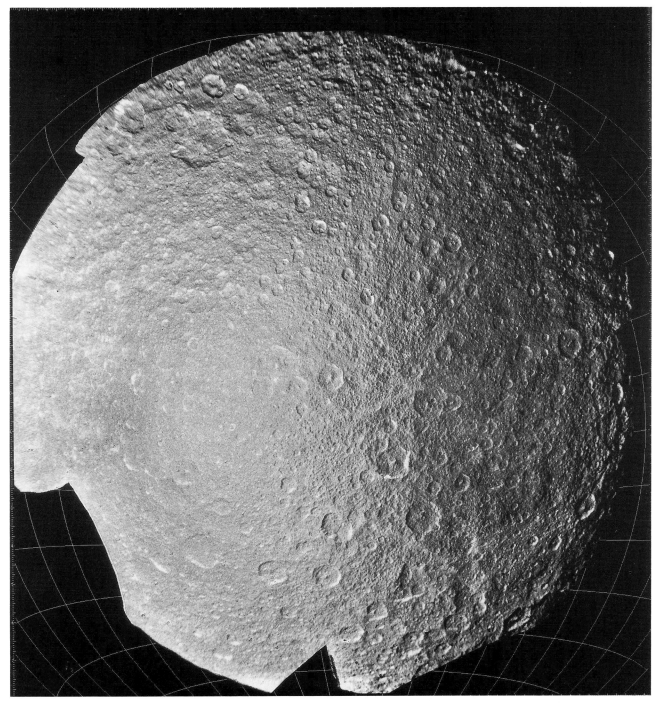

Figure 9.23 Mosaic of Rhea showing the subsaturnian hemisphere along the equatorial band and resurfaced terrain along the terminator in the northern hemisphere.

Figure 9.24   Shaded airbrush relief map of Rhea (courtesy US Geological Survey).

Sub-saturnian hemisphere

Figure 9.25 Physiographic map of Rhea (from Moore *et al.* 1984).

Richard Pike and Paul Spudis (personal communication) have discovered a second, very degraded basin centered at ~ 30°N, 315°W and more than twice the size of the basin discussed by Moore *et al.* (1985). The center of the basin is in the resurfaced terrain. Subsequent impact cratering of this region has muted the topography, rendering the inner rings into subtle ridges with gentle slopes.

Some resurfaced terrains, such as those centered at 25°N, 310°W, may represent flooding by water 'magmas'. Parallel, N 30°E-striking, sharp-crested ridges or scarps occur in resurfaced terrain between 270°W and 360°W longitude, especially near the crater Pedn. They are ~ 5 km wide, up to several hundred kilometers long, and their relief averages < 0.25 km. Some ~ 20 km craters are superimposed on the ridges or scarps, whereas other craters are transected by ridges. The rim of Pedn shares a common scarp with two lineaments on its northwest and southeast extensions, implying a common subsurface control of the scarps or ridges and the crater.

Parallel linear troughs or coalescing pits oriented at N45°E occur in older cratered terrain between 0° to 60°N and 60° to 120°W. These features are ~ 5–10 km wide and up to ~ 1000 km long. Shadow measurements indicate a depth of < 0.5 km. Nearly all troughs and many of the coalescing pits show little or no indication of raised rims. Some of the pits have raised rims, implying either an impact origin or features that were sites of pyroclastic activity. Most troughs and pits are superposed on older impact craters. As they cannot be traced into younger cratered terrain, Moore and Horner (1984) infer that

trough and coalescing pit formation continued beyond the end of heavy bombardment but ended prior to the resurfacing event which mantled the older craters. The troughs and coalescing pits may, therefore, be a consequence of an episode of extensional tectonics and volcanism occurring between the end of heavy impact bombardment and the resurfacing event.

9.2.6 Titan

Titan is a remarkable satellite. Exceeding the diameter of Mercury, at 5120 km it is in the same size class as Ganymede and Callisto and, like those satellites of Jupiter, its mass suggests a composition that is about 45 percent water ice and 55 percent rocky material. What makes it remarkable is the presence of a dense atmosphere. Although the satellite was the first to be discovered in the saturnian system (found by Huygens in 1655), the presence of an atmosphere was not suspected until the mid-1900s (as reviewed by Owen 1983). In 1944, the Dutch–American astronomer, Gerard Kuiper, identified methane in the spectrum of Titan; this initiated speculations on the various chemical reactions that might occur in the atmosphere.

The presence of extensive clouds on Titan was well established by the time the Voyagers arrived. Voyager investigators hoped that the clouds would be broken and sufficiently scattered to allow glimpses of the surface. Such was not the case, and the Voyager pictures are

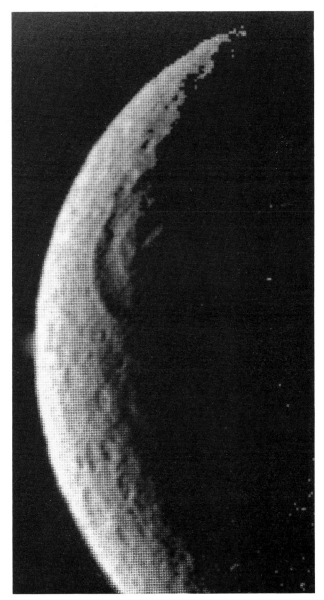

Figure 9.26 Multi-ringed basin about 450 km across found on Rhea (Voyager 1 1152S1 + 009).

Figure 9.27 Voyager image of Titan showing structure within the atmosphere considered to be photochemical hazes (Voyager 2 JPL P-24067).

disappointingly bland, showing a relatively uniform orange sphere with only hints of variations in the clouds. Limb views, however, show layers in the atmosphere (Fig. 9.27), which are thought to be smog-like photochemical hazes. It is likely that very little sunlight penetrates the clouds, and the surface must be a frigid, gloomy place.

The non-imaging experiments on board Voyager yielded results that show Titan to be a fascinating object. Nitrogen was found to be the principal component in the atmosphere, and Voyager also discovered ethane, acetylene, ethylene, hydrogen cyanide and various other carbon–nitrogen components. It is thought that when Titan first formed, methane and ammonia ices were incorporated into the accreting mass. With heating and subse-

quent differentiation, these compounds were released from the interior and formed an early-stage atmosphere. The action of sunlight on the ammonia caused release of the nitrogen and escape of most of the hydrogen, while methane remained as a planet-wide ocean. In a 1983 conference, however, Lunine, Stevenson and Yung (personal communication) showed that methane which evaporates from the surface of this ocean would be photodissociated into compounds which would then recombine to form ethane and acetylene. These and other molecules would then condense and rain down onto Titan's surface.

Surface temperatures are about $-180°C$, with perhaps a $3°C$ variation from the equator to the poles. This temperature, however, is near the triple point of ethane and methane, and it has been suggested that Titan may be covered with a liquid ethane–methane ocean to a depth of 1 km, which might have islands of solid H_2O ice rising from the ocean floor. Liquid ethane and methane may play the same role on Titan as water does on Earth, so far as surface processes are concerned. Sagan and Dermott (1982) have calculated that the present eccentricity of Titan's orbit raises large ($\sim$ 10 m high) tides in a global ocean that quickly erodes any surface above ocean-level, suggesting that if there is dry land, it would have to be newly formed.

We can only speculate what geologic events might take place on Titan, either at present or in the past. Fluid processes may modify solid parts of the surface and could involve rivers of liquid ethane–methane, tidal activity by

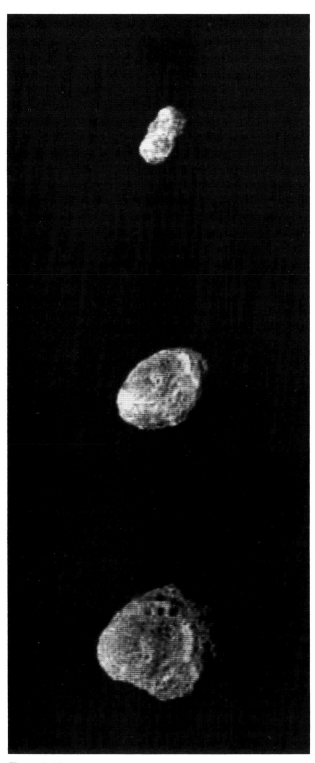

Figure 9.28 Voyager views of Hyperion showing irregular shape and the presence of large degraded craters and crater-related scarps (Voyager 2 JPL P-23932).

an ocean, and perhaps even aeolian processes involving windblown ice particles. Whether these or other geologic processes, such as volcanism, tectonics or impact cratering, have occurred and left landforms is open to question. Similar to the situation for cloud-covered Venus, a radar-imaging system is required to view the surface of Titan; such a system has been discussed for some future spacecraft mission. In many respects, Titan resembles a primordial Earth, and the presence of organic compounds in a chemically rich atmosphere makes Titan a target of high priority, not only for atmospheric and geologic sciences, but for biology as well.

9.2.7 Hyperion

Hyperion is a small, irregularly shaped object whose elliptical orbit lies just outside that of Titan. Measuring 350 by 235 by 200 km, its shape includes both angular and rounded parts. Hyperion is the darkest object of the inner saturnian satellites and, although its mass cannot be estimated from Voyager data, it is presumed to be composed of a mixture of ice and rock.

Hyperion is heavily cratered. Large craters—some as large as 120 km in diameter with 10 km of relief—are scattered over the surface, while craters 10 km across are relatively abundant (Fig. 9.28). In addition, the surface displays segments of scarps which may be part of a connected, arcuate feature nearly 300 km long. This feature could be the remnant of a craterform that would have a diameter of > 200 km.

Because of its irregular shape, heavily impacted surface and the possibility of an impact scar nearly half its size, Hyperion is considered by most investigators to be the remnant of a much larger object which was shattered by impact processes.

9.2.8 Iapetus

Iapetus is nearly the twin of Rhea in its diameter, but its mass suggests a much lower density, about 1.1 g cm^{-3}, and a higher proportion of ice than for Rhea. Since the satellite's discovery in the 17th century by Cassini, the albedo contrasts on Iapetus have long been recognized as distinctive. While the leading hemisphere is very dark, the trailing hemisphere is about an order of magnitude brighter and, as found on Voyager images, the boundary is fairly sharply defined (Fig. 9.29).

Surface features are difficult to see in the dark area (named Cassini Regio), but craters up to 120 km in diameter are clearly visible in the north polar area and in the bright terrain of the trailing hemisphere (Fig. 9.30), despite the relatively poor resolution of the images (at best, ~ 10 km). The large craters display central peaks and have well-defined rims. Some of the craters in the bright terrain along the boundary with the dark terrain have floors covered with material that is dark and that has the same spectral properties as the dark terrain. In addition, a dark ring about 200 km across, which may be a multi-ring impact basin (Smith *et al.* 1982), was imaged by Voyager 1 (Fig. 9.31).

The dark material is about as black as tar or asphalt; its albedo is 3–5 percent. The only known materials of this darkness common in the Solar System are carbonaceous chondrites. Contact relations and the possible filling of some craters suggest that the dark material is a mantle, or surficial deposit. The origin of the dark material on Iapetus is somewhat controversial, and two general ideas, each with variations, have been proposed. One model suggests internal sources (eruptions from the interior) and the other suggests origins by external sources. In either case, most investigators agree that the material seems to be a mantling deposit that is draped over bright terrain, as inferred from contact relations.

Because of the symmetry of the dark material in relation to the leading hemisphere, even before the Voyager mission Cook and Franklin (1970) proposed that impacts on the leading hemisphere could erode the ice components in the lithosphere and leave a lag deposit of dark, rocky materials. This idea requires a high proportion of rocky materials to ice, which does not fit the overall low density of the satellite. Steve Soter of Cornell University suggested at a conference in 1974 that Phoebe, the satellite in the next orbit outward from Saturn and a very dark object itself, was the source of dark material on Iapetus. Bombardment of Phoebe by impact cratering would eject debris which would fall inward toward Saturn. He suggested that some of this material would intersect the orbit of Iapetus and would be deposited on the leading hemisphere. The main weakness with this argument is that the spectral properties of Phoebe and the dark terrain on Iapetus do not match (Iapetus is redder), suggesting that they are not the same material.

An internal origin for the dark material was proposed by Smith *et al.* (1982). They observed that the bright terrain appears to be flooded by dark material because dark material occurs in topographically low areas, such as crater floors. In their model, materials were erupted onto the surface from the interior, perhaps as a carbonaceous-rich or organic-rich ice slurry.

Despite the additional information provided by the Voyagers, the origin of the dark material on Iapetus remains open to speculation.

9.2.9 Phoebe

Phoebe, the outermost saturnian satellite, is in a retrograde orbit inclined to the equatorial plane of Saturn. This atypical orbit suggests that Phoebe may be a captured object. Estimates of the mass of this ~ 220 km size satellite could not be made by Voyager because the spacecraft did not pass close enough to the satellite; however, its low albedo and general spectral properties suggest a surface consisting of carbonaceous material. Voyager images are very poor—about 20 km resolution at best—but show Phoebe to be approximately spherical and the darkest object in the saturnian system. Circular markings and indentations suggest the presence of craters.

Analysis of the Voyager color data and ground-based multicolor photometric results suggested to Voyager team members (reviewed in Smith *et al.*, 1982) that Phoebe is similar to one class of asteroids that are common in the outer Solar System. Thus, Phoebe may be the first unmodified primitive object to be seen in the outer Solar System by a spacecraft.

9.2.10 Small satellites

In addition to the satellites described above, there are eight additional known satellites. All are rather small, ranging in size from 25 to 220 km (Fig. 9.32). Several have been officially named, but all of them still carry the designation that indicates the year of discovery (all were found in 1980, some of them from Voyager images).

Five fall within the orbit of Mimas (Fig. 9.1). Two (Janus, 1980S1, and Epimetheus, 1980S3) are called the co-orbital satellites because their orbits are within 50 km of each other. They are thought to be remnants of a larger object which was fragmented by an impact.

Two satellites, 1980S26 and 1980S27, are called "ring shepherds" because they bracket the F Ring and are thought to confine the orbits of the particles that make up the ring. The A Ring shepherd, Atlas (1980S28), is about 30 km across and one of the smallest known saturnian satellites.

The remaining three satellites, Calypso (1980S13), Telesto (1980S25) and the unnamed 1980S6, are known as the Lagrangian satellites. As reviewed by Morrison (1982b), it has long been known that a small object can have the same orbit and orbital speed as a larger object, so long as it maintains 60° of arc ahead or behind the larger object (these positions are called the Lagrangian points). Tethys has four such small objects in its orbit, including Calypso and Telesto; Dione has one, as shown in Figure 9.1.

9.3 Geologic processes

Collectively, the surfaces of the saturnian satellites display landforms indicative of impact cratering, tectonic deformation and processes associated with the release of fluids from the interior. Direct evidence for gradation is mostly lacking, probably owing to insufficient image

Figure 9.29 Shaded airbrush relief map of Iapetus (courtesy US Geological Survey).

246

Anti-saturnian hemisphere

Figure 9.30 Voyager 1 view of the north polar region of Iapetus showing portions of both the bright and the dark hemispheres. The ring of dark material may indicate the presence of a large impact basin (Voyager 2 JPL P-23920).

Figure 9.31 Voyager 1 view of Iapetus showing the sharp contact between the two albedo units and the apparent "flooding" of crater floors with dark material (Voyager 1 JPL P-23105).

Figure 9.32 Mosaic of Voyager images showing some of the small saturnian satellites; upper left: 1980S26 (outer F-ring shepherd), 120 by 100 km; lower left: 1980S27 (inner F-ring shepherd), 145 by 70 km; upper right: 1980S1 (Janus), 220 by 160 km; lower right: 1980S3 (Epimetheus), 140 by 100 km (Voyager 2 JPL P-24061).

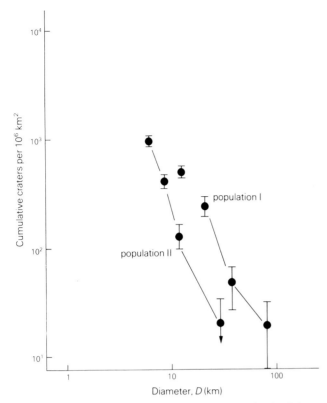

Figure 9.33 Impact crater frequency distributions for the light cratered terrain and heavily cratered terrain on Dione, difference reflecting population I (HC) and population II (LC) craters (modified from Plescia & Boyce 1982; reprinted by permission from *Nature* **295**, 3; copyright © 1982 Macmillan Journals Ltd).

resolution; however, mass wasting and impact erosion on a wide range of scales almost certainly occur now and have occurred in the past.

9.3.1 Impact cratering

As with the Galilean satellites, Ganymede and Callisto, impacts on the satellites of Saturn involve ice-rich targets. However, because the saturnian satellites are smaller and less dense (except for Titan), their internal temperatures will be much lower (Table 2.1), and the ice, because of its lower temperature, may behave more like rock, both during impact and in the post-impact modification stages. Furthermore, the relatively smaller sizes of the saturnian satellites means that gravitationally driven viscous relaxation of topography would be less pronounced than on Ganymede and Callisto.

Most of the craters on the saturnian satellites are circular to polygonal in outline, are generally bowl-shaped, and have more or less lunar-like depth-to-diameter ratios. However, the 400 km Odysseus crater on Tethys, the two unnamed multi-ring basins on Rhea, and some of the craters on Enceladus appear to be topographically "relaxed". These craters are generally ancient in compari-

son to the other features on the respective satellites, and it is thought that they may have formed at times when the temperatures of the objects were higher, i.e. following the stage of accretion when impact-generated heat may have been retained.

Smith *et al.* (1981; 1982) analyzed impact crater distributions on the saturnian satellites and compared them with present-day fluxes both near Saturn and throughout the rest of the Solar System. As a starting point, they considered the accepted variation in impact cratering within the Saturn system, taking into account the differences in gravitational focusing by Saturn in relation to the orbital distance for each satellite, as well as differences in mass among the satellites and the effect on impact energies (and hence differences in crater size).

Although there are uncertainties in the modeling, Smith *et al.* (1982) note that Iapetus and Phoebe would be least cratered, and that the inner satellites could be more heavily cratered by two orders of magnitude! For example, at the current rates of bombardment by comets (the primary source of bolides in the outer Solar System), Rhea would have twice the number of craters as Iapetus, while Mimas would have 20 times the craters. Moreover, there could be a ratio as large as 20:1 in the crater frequency from the leading hemisphere to the trailing hemisphere for the satellites.

From the considerations outlined above, Smith *et al.* (1982) recognized two distinct sets of craters on the saturnian satellites (Fig. 9.33), which they termed population I and population II craters. Population I terrain is characterized by abundant craters > 200 km in diameter which are considered to represent the end of the period of heavy bombardment, approximately equivalent to the heavily cratered terrain of the inner planets.

Population II terrain has a high frequency of craters < 20 km in diameter and relatively few large craters. The size–frequency of the population II craters is statistically similar to the distribution of secondary crater populations on terrestrial planets. Analyses by Smith *et al.* (1982) and Plescia and Boyce (1982; 1983) show that both populations are present on Dione, Rhea and Tethys, while Mimas is dominated by population II, craters, and Enceladus lacks population I craters entirely, evidently reflecting differences in resurfacing histories.

One of the more intriguing discoveries to come from the Voyager mission is the possibility that the satellites have been repeatedly disrupted and reaccreted. Smith *et al.* (1981; 1982) estimate that all of the satellites from Dione inward have been struck at least once by a body with sufficient kinetic energy to form a crater equal in size to the satellite being struck. On the inner satellites, such impacts would have occurred many times. Thus, statistically, it is likely that the satellites have been broken apart repeatedly. They note that, because of the circular orbits, most of the disrupted material would accrete to form a reassembled body. This may explain the absence of population I craters on Mimas.

9.3.2 Tectonism and volcanism

Prior to the Voyager flybys of Jupiter and Saturn, models of satellite evolution by Lewis (1971) and other early studies suggested that the larger, denser satellites would experience heating by impact cratering following accretion and from radionuclides, leading to chemical differentiation in which the rocky materials would form cores, while volatiles (mostly water) would form mantles and lithospheres. Smaller or less dense objects were thought not to experience sufficient heating to lead to differentiation but would remain as cold, homogeneous mixtures of ice and rocky material. Lewis, however, noted that ammonia may well be present in the outer Solar System, and its addition could lower the melting point for the mixture.

Observations of the surface features and the inferred histories for the saturnian satellites have caused reassessment of the evolutionary models for small, volatile-rich satellites. Volcanism involving icy slushes and tectonism (Poirier 1982) that has fractured the moons, in some cases on a grand scale (Fig. 9.17), requires explanations that take into account not only the sizes, densities and relative positions of the satellites in the saturnian system, but also the very origin of the Solar System, as reviewed by Smith et al. (1981; 1982).

Ellsworth and Schubert (1983) note that the four largest satellites (excluding Titan), Tethys, Dione, Rhea and Iapetus, are large enough and contain sufficient heat-producing rocky materials to have generated convection beneath a lithosphere. This process is independent of initial thermal conditions and could have generated the observed surface features. Mimas is so small that any thermal activity would have occurred early in its history due to accretionary processes. In contrast, Enceladus must experience heating from tidal stresses, although estimates by Squyres et al. (1983) based on present conditions indicate that these are inadequate to melt the interior.

Stevenson (1982) has assessed the role of volcanism and igneous processes on the saturnian satellites and concludes that an ammonia–water magma is required to explain the resurfacing on Enceladus. He suggests that explosive venting could occur if clathrates are involved with the heating on all the larger satellites. Contact between the magma and one or more clathrates of methane, nitrogen and argon could cause pyroclast-forming explosions to generate features such as the bright, wispy albedo patterns seen on Dione and Rhea.

9.4 The future

Besides providing many important discoveries, the Voyager mission raised a great many questions for the saturnian system. From the organic "soup" of Titan to the curious "shepherding" satellites of the rings, the Saturn system remains of high intrinsic interest to planetary science. The Cassini mission, planned for early in the 21st century, is a Galileo-like mission involving an orbiter and various probes to provide many answers to questions raised by Voyager.

10 The Uranus system

10.1 Introduction

Travelling in excess of 20 km s^{-1} in early 1986, the Voyager 2 spacecraft streaked past Uranus on its continuing journey through the outer Solar System. In its brief encounter, Voyager 2 returned more information on Uranus, its rings and its satellites than had been accumulated in the 205 years since the discovery of this, the seventh planet from the Sun. With the return of these data and the discovery of ten new satellites, the Uranus system can now be incorporated into the family of objects available for geologic study. An excellent collection of papers on the history of studies of Uranus and of pre-Voyager data is given by Hunt (1982).

Uranus was the first planet to be found in historical times. Its discoverer, William Herschel, earned his living as a musician in Bath, England, but was also an amateur astronomer who enjoyed building telescopes for observing stars. In March 1781, while using a 7-foot (~ 2 m) telescope to search for double stars, Herschel found a curious object that did not behave like a normal star. Although first described as some sort of nebular star or a comet, additional observations revealed the object to be a planet. Herschel wanted to name the planet "George" after the reigning King George III, but instead, Bode suggested that the newly found object be named Uranus after the Roman god of the heavens (Smith 1982). William Herschel continued viewing Uranus and in 1787 he discovered its two largest satellites, later named Titania and Oberon. Another amateur British astronomer, William Laswell, found two additional satellites, Umbriel and Ariel, in 1851. The fifth large satellite, Miranda, was not discovered until nearly 100 years later, by the well-known planetary astronomer, Gerard Kuiper.

The uranium rings were found by accident in 1977. A team led by Jim Elliot was observing star occultations by Uranus, using NASA's special airborne observatory. They noted several occultations immediately before and after the passage of Uranus in front of the star and—with subsequent observations—confirmed the existence of nine thin, dark rings encircling Uranus. Beginning with the outermost ring, they are designated epsilon, delta, gamma, eta, beta, alpha, 4, 5 and 6 (the innermost ring).

Uranus is the third largest planet in the Solar System, after Jupiter and Saturn. Like those two giants, Uranus is composed mostly of hydrogen and helium. However, it has a slightly higher density, suggesting the presence of heavier elements which are probably contained in a silicate-rich core hidden below a deep "ocean" and dense atmosphere. Viewed telescopically and in color-corrected spacecraft images, the atmosphere of Uranus has a distinctive blue-green appearance due to the presence of methane which absorbs red light.

Unlike any other planet, the axis of rotation for Uranus is tipped 82° from the vertical, so that it lies nearly horizontal on the ecliptic plane. Although no wholly satisfactory answer has been found to explain this orientation, a plausible hypothesis suggests that Uranus was struck early in its history by an Earth-size object which knocked it askew. At present, the south pole points towards the Sun. Thus, as the rings and satellites rotate about the equator, their paths inscribe patterns similar to a bull's eye.

Most of the information on Uranus and the geology of the satellites discussed in this chapter has come from the Voyager mission (Smith *et al.* 1986). Acquisition of the Voyager data presented some new problems in planetary exploration. At a distance of 5 billion kilometers from the Sun, the amount of light reaching Uranus is very low, with only about 1/400 the level of light that reaches Earth. This means that the camera exposures had to be as long as 96 s in order to record an image. Because the spacecraft was traveling so fast, these long exposures would have resulted in 'smeared' pictures had not an **image motion compensation** (*IMC*) routine been introduced to the spacecraft operating system. IMC involves rotating the spacecraft during the exposure so that the camera remains fixed on its target. This capability and means for 'data compression' (extracting the most useful data before transmission to Earth) were not included in the mission design when Voyager 2 was launched in 1977. But, through the skill of the engineers at the Jet Propulsion Laboratory and the foresight to be able to reprogram the spacecraft in flight, these critical additions to the mission permitted the return of more than 7000 images and other useful data.

Review papers written on the Uranus system in conjunction with a symposium are given in Bergstrahl *et al.* (1991).

10.2 Uranus the planet

Even before the Voyager 2 encounter, models of the structure of Uranus had been formulated based on cosmic abundances of the elements and general theory of Solar System evolution. The planet can be divided into three regions—a rocky core, an "ocean" and a thick atmospheric envelope. Spacecraft data have allowed further refinement and development of these models, and have revealed some surprises. Prior to the Voyager flyby, it was predicted that Uranus would have a magnetic field. The magnetometer on board the spacecraft verified the presence of a strong field—comparable to that of Saturn and Earth—but also showed, unexpectedly, that the orientation of the field is offset from the axis of rotation by some 60°. For comparison, Earth's magnetic field is offset by only 11°. Perhaps, like Earth, Uranus experiences periodic reversals of the north and south magnetic poles and such a reversal may be occurring now. Alternatively, if Uranus was struck early in its history and its spin axis reoriented, the magnetic field orientation may be a remnant from that time. In any event, the magnetic field indicates either a fluid within the core similar to the "dynamo" model suggested for the generation of Earth's magnetic field, or some other arrangement. One model, proposed by David Stevenson of the California Institute of Technology, relates the field to the uranium hydrogen ocean. The ocean is postulated to be thousands of kilometers thick and, at depth, the high pressures would raise the temperature in excess of 1000 K. At these temperatures, water would dissociate into hydronium and hydroxide ions which would conduct electricity, thus permitting the generation of a magnetic field as the planet rotates.

The atmosphere of Uranus has long been of interest to planetary scientists. Composed mostly of hydrogen, a key question prior to the Voyager encounter was the amount of helium present. Although predictions ran as high as 40 percent, Voyager showed that only about 12 percent is present. In addition to hydrogen and helium, small amounts of methane, ammonia and other gases are present. In unprocessed images, the atmosphere appears quite bland and lacking in the rich cloud patterns seen on Jupiter. Computer enhancement of the images, however, reveals a brownish haze in the south polar region (the north pole, pointing away from the Sun, was not observed). The haze may be a complex organic compound, perhaps produced by solar radiation altering methane and other materials in the atmosphere.

Image processing also shows several white, cloud-like features in the upper atmosphere and faint, but distinctive, latitudinal bands indicative of lateral shear. These features were tracked for several days and enabled detection of upper atmosphere high-speed winds. The winds move in the same direction as the rotation of the planet, a result that was just the opposite of predictions. Thus, despite its bland appearance, the atmosphere of Uranus has posed a number of intriguing questions not yet answered.

10.3 Rings

Several Voyager observations were designed to gather data on the rings. The mission objectives included viewing the known rings, searching for new rings and searching for "shepherd" satellites like those known to accompany some of the saturnian rings. As was known from Earth-based observations, Voyager showed the rings to be very dark—having an albedo of only ~ 5 percent (similar to soot, or lampblack)—and a color that is grey. This is in sharp contrast to the rings of Jupiter and Saturn, which are much brighter and have a reddish coloration, and it has been suggested that the uranian rings may be composed of carbonaceous materials. In addition, the proportional amount of micron-size dust in the main rings is orders of magnitude less than in rings of Saturn and Jupiter, suggesting different processes of particle interactions. At least two new, complete rings and ten additional rings or partial rings (ring "arcs") were also found.

The epsilon ring ranges from ~ 30 to 100 km in width. It appears to be composed of two "strands" of particles separated by a dark region and shows complex patterns. Although an intense search was conducted throughout the ring system, only the epsilon ring appears to have shepherding satellites—designated 1986U8 for the outside satellite and 1986U7 for the inside satellite. Voyager also discovered a broad band of dust inside rings 4, 5 and 6. The band is very faint and extends to within 11,000 km of Uranus. This dust band, the distribution of particle sizes in the main rings, and the presence of possible incomplete rings, or "arcs", all suggest that the rings are in an evolving state. Perhaps repeated impacts cause fragmentation of small moons, with the fragments being swept up later or incorporated into newly formed and "incomplete" rings.

10.4 Satellites

The five largest moons, Miranda, Ariel, Umbriel, Titania and Oberon (Table 2.1) are named after characters in Shakespeare's *A Midsummer Night's Dream* and *The Tempest*, and Pope's *The Rape of the Lock*. All the satellites are in synchronous rotation, are dark, and have densities of ~ 1.5–1.7 g cm^{-3} (Table 2.1). As shown in Figure 10.1, these densities are less than the Galilean satellite, Ganymede, but are slightly higher than the saturnian satellites, suggestive of a higher proportion of rock to ice.

Voyager images show that all five satellites have experienced impact cratering, tectonic deformation and resur-

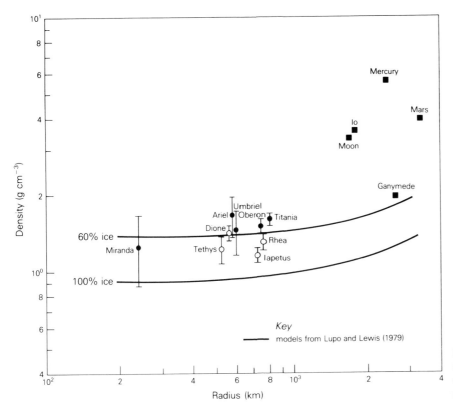

Figure 10.1 Density versus radius for the five largest uranian satellites compared with other satellites and planets (from Smith *et al.* 1986).

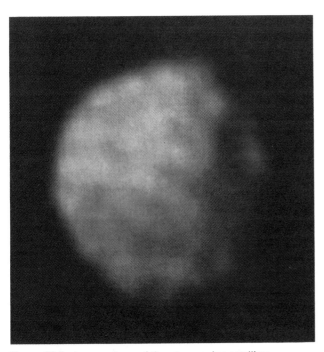

Figure 10.2 Image of one of the new uranian satellites discovered during the flyby of Voyager 2. Designated 1985U1, this satellite is about 150 km in diameter and shows numerous craters (Voyager 2 JPL P-29519).

facing, although the intensity of these processes varies among the satellites. The outermost moons seem to have experienced little modification since the period of heavy impact bombardment, whereas the innermost satellites— Miranda and Ariel—have been extensively modified by internal processes.

Ten new moons were discovered through Voyager. Not yet formally named, they are designated 1985U1 (Fig. 10.2) and 1986U1 through U9. All are small (30–180 km in diameter), reside within the orbit of Miranda and—like their larger neighbors and rings—have low albedos. The moons' masses are difficult to determine and the data from the Voyager mission are inadequate to allow estimates of their densities to be obtained. However, it is presumed that the densities of the newly discovered satellites are similar to those of their larger neighbors.

Reflectance spectroscopy data suggest that water ice is present on all the satellite surfaces but, like the rings, the moons are rather dark and grey in color. This implies either that they are composed of dark material, or that they are mantled by dark material which has been swept up from orbit. Alternatively, a model has been proposed (Calcogno *et al.* 1985; Cruickshank *et al.* 1983) to explain the dark surfaces seen on many of the outer planet satellites, including those of Uranus. Through laboratory experiments, it has been shown that protons impacting methane ice for a prolonged period drive out hydrogen molecules, leaving a high proportion of carbon at the

Figure 10.3 View of Oberon, showing its heavily cratered surface. Light areas around craters represent ejecta deposits and include bright-ray material (Voyager 2 JPL P-29501).

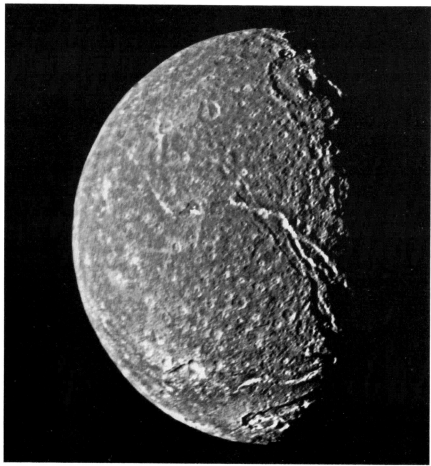

Figure 10.4 Titania, seen at a resolution of 3.5 km, shows impact craters, multi-ring basins (upper right) and grabens representing tectonic processes (Voyager 2 JPL P-29522).

crust, perhaps in response to expansion which occurred as ice froze in the interior of the satellite. Smooth, relatively uncratered zones appear to be areas resurfaced by extrusion of fluid-like materials.

Umbriel is the darkest of the large satellites and appears to be quite uniform in its color, albedo and general surface features, as shown in Figure 10.5. Although Voyager images do not cover all of Umbriel, that part observed is uniformly cratered with a high proportion of large impact scars, indicative of an ancient surface. Although it is heavily cratered, ejecta rays are not visible, nor are other small-scale surface features, suggesting to Smith *et al.* (1986) that Umbriel has been blanketed with dark material from external sources. Alternatively, the material excavated during impact cratering may be of a composition whose optical properties are unaffected by impact and hence does not form bright ejecta rays and deposits.

The geomorphic features revealed by Voyager for Ariel

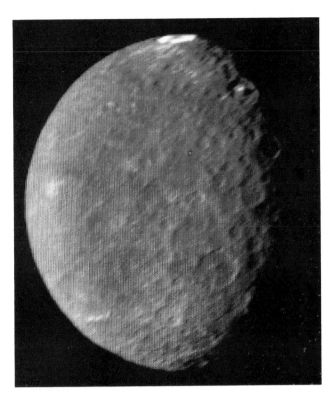

Figure 10.5 Umbriel, the darkest of the uranian satellites, is about 1200 km in diameter and shows little geologic activity other than impact cratering. The prominent crater in the upper right is about 110 km across and has a large central peak. The bright ring at the top of this view of the southern hemisphere may be a frost deposit (Voyager 2 JPL P-29521).

surface, which causes a general darkening. Although it has not been shown that sufficient methane ice is necessarily present in the ring and satellite materials nor that such irradiation would occur in the uranian system, it is a viable hypothesis to explain the low albedos.

Oberon, the outermost satellite, is the second largest moon in the uranian system. Its surface is dominated by large (50–100 km) impact craters, some of which exhibit radial patterns of bright ejecta (Fig. 10.3). Faint, arc-like features suggest remnants of impact basins. Some linear and curved scarps appear to be faults. As shown on a geologic sketch map by Smith and colleagues (1986), several dark patches occur on Oberon, particularly on the floors of large craters. These areas are thought to represent the extrusion of material to the surface from the interior following the period of heavy impact cratering.

With a diameter of 1610 km, Titania is the largest satellite of Uranus (Fig. 10.4). Like Oberon, much of its surface is heavily cratered and includes multi-ring basins. However, numerous faults, including features 20–50 km wide which appear to be grabens, suggest more extensive tectonic deformation than occurred on Oberon. High-albedo materials—possibly ice or frost deposits—are exposed on some of the scarps. Smith *et al.* (1986) suggest that the faults represent global extension of Titania's

Figure 10.6 Mosaic of high-resolution (1.2 km/pixel) images of Ariel, showing its cratered and fractured surface (Voyager 2 JPL P-29520).

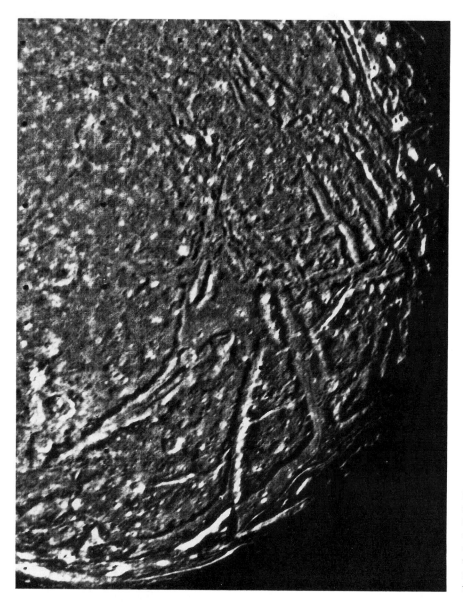

Figure 10.7 Close-up of Ariel showing terrain near the terminator. Complex graben system indicates extensive tectonic modification; many of the grabens appear to be floored with smooth deposits which may represent extrusion of material from the interior (Voyager 2 JPL P-29523).

(Fig. 10.6) are much more complex than those seen on the outer, larger satellites. The older, cratered terrain on Ariel has been extensively fractured by faults and grabens (Fig. 10.7). However, the population of very old, large craters appears to be lacking, perhaps having been obliterated by an early episode of resurfacing. Smaller (< 60 km in diameter) craters are present in some areas; other areas consist of smooth plains, some of which may have been emplaced as flows which buried the older terrain. Some of the grabens on Ariel show sinuous valleys on their floors and the Voyager team raises the question of whether these features represent fluid flows or are sinuous faults.

Although Miranda is the smallest of the major satellites, it displays the most complex geologic history (Fig. 10.8). As shown on a preliminary sketch map, various terrains can be identified on Miranda: an old, heavily cratered terrain having a "rolling" or gently undulating surface of uniform albedo, and younger, complex terrains. The younger terrains occur in at least three areas, each of which has been given informal names by the Voyager imaging science team based on the general shape and appearance (Smith *et al.* 1986): (a) the "trapezoid", (b) the "banded ovoid" and (c) the "ridged ovoid". These younger units show complex patterns of grooves and ridges (Fig. 10.9), similar in many respects to the tectonically deformed bright terrain on the Galilean satellite, Ganymede (Fig. 8.28) and the saturnian satellite, Enceladus (Fig. 9.6). Patches of smooth, uniform-albedo material within the complex terrain are suggestive of resurfacing by fluids.

Miranda also displays enormous fault scarps (Fig. 10.10), some of which cut across the entire globe. The largest of these include grabens 10–15 km deep. Bright material is exposed in the walls of some fault scarps and craters. Extending perhaps 1 km below the surface, this

256

Figure 10.8a Mosaic of high-resolution (300 m/pixel) images of Miranda in an orthographic projection, centered on the south pole (courtesy R. Batson, US Geological Survey, Flagstaff).

may be a zone of relatively "clean" ice that has been exposed by faulting and crater excavation.

10.5 Geologic processes

The record of impact cratering exhibited on planetary surfaces is a fundamental key to deriving geologic histo-

ries. The Voyager imaging science team (Smith *et al.* 1981; 1982) suggested that different "populations" of craters could be recognized on the saturnian satellites (see Section 9.3.1) and this concept has been extended to the satellites of Uranus.

Population I craters represent post-accretional heavy impact cratering and dominate the surfaces of Oberon and Umbriel. Population II craters have a lower frequency of large craters and are prevalent on Titania and Ariel, although some large craters are also present. Miranda also shows both populations, with its old, heavily cratered terrain showing more craters ≤ 50 km than any surface on the other four satellites. The differences in the cratering record among the satellites can be explained by differences in tectonic deformation and resurfacing. Both Oberon and Umbriel appear to have been little modified following the time of post-accretional heavy bombardment, whereas the crusts of Titania and Ariel have been extensively deformed and resurfaced by extrusive materials.

Despite their small sizes, Miranda and Ariel clearly have experienced substantial modification by internally generated processes. Extensive tectonism and volcanism

suggest internal heat sources which have caused a reassessment of the mechanisms responsible for heat generation. Prior to the Voyager discoveries, the conventional 'model' suggested that small objects would have a low amount of radionuclides in the interior and hence a low heat supply; therefore, they should exhibit few surface features indicative of volcanism and tectonism. However, as discussed in Chapter 9, the Voyager results have changed this perspective. These and other results from the Voyager mission will continue to stimulate planetary geology. However, as outlined in Chapter 11, there are no plans to return to Uranus in the 20th century, and the Voyager 2 data will be the primary source for information on the uranian satellites for a long time.

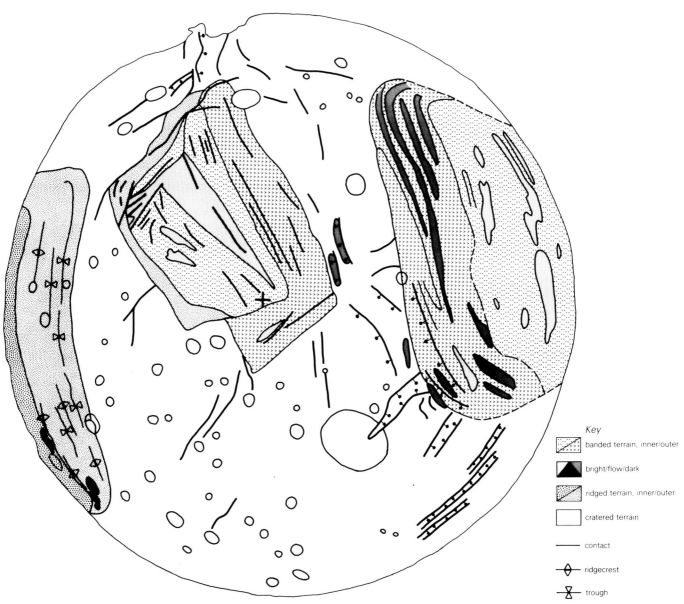

Figure 10.8b Geological sketch map of Miranda, corresponding to the mosaic of Figure 10.8a (from Smith *et al.* 1986).

Figure 10.9 High-resolution image of Miranda showing cratered terrain (center), banded terrain (upper right), and ridged terrain (lower left). Area shown is about 220 km across (JPL P-29515).

Figure 10.10 View of Miranda, showing deeply incised grabens. Area shown is about 180 km across (JPL P-29513).

11 The Neptune System

11.1 Introduction

The existence of Neptune, outermost of the giant gaseous planets, was predicted before it was actually discovered. Telescopic observations of Uranus in the early 1800s showed notable discrepancies between its predicted position in orbit and where it was actually observed. Independently of each other, two mathematicians, John Couch Adams of Cambridge, England and the Frenchman Urbain John-Joseph Le Verrier, analyzed the problem and suggested both the position and the mass of an as-then unknown planet to account for the abnormal motions of Uranus.

Adams was essentially ignored by the English scientific establishment, while Le Verrier—although gaining the attention of the French Academy—was unable to convince French astronomers to search for the possible new planet. He did, however, stimulate the interests of a young German astronomer, Johann Gottfried Galle, who used the prime telescope of the Berlin Observatory to look for the predicted planet. On the very first night of searching by Galle, September 23, 1846, the eighth planet was found within 1° of its predicted position.

Only 17 days after the discovery of Neptune, its largest moon, Triton, was found by William Laswell. But it would be more than a century before Neptune's other large satellite, Nereid, would be discovered by Gerard Kuiper in 1949, and still later, the mid-1980s, before Neptune's ring-arcs were found.

In its final phase of Solar System exploration, Voyager 2 began observing the Neptune system in June 1989, as reported by Stone and Miner (1989). Within a few days, more information on the Neptune system was returned than had been accumulated in more than 14 decades of observations. The camera system acquired more than 9000 images, which were analyzed by Smith and the Voyager team (1989). Knowledge was gained on the magnetosphere and atmosphere of Neptune, its ring-arcs were defined, six new satellites were discovered, and, in a spectacular closing to what can only be described as a fantastic journey of discovery, Voyager returned images of actively erupting geysers on Triton.

11.2 Neptune the planet

After its discovery, subsequent telescopic observations showed Neptune to have a nearly circular orbit. Because of the enormous distances in the outer Solar System, 165 years are required for Neptune to complete one trip around the Sun. The planet is about three times smaller than Jupiter and, like Uranus, it is probably composed mostly of hydrogen and helium.

Neptune is the densest of the jovian planets, suggesting that heavy elements are contained in the interior. The planet emits some 2.7 times more energy than it receives from the Sun, and has a strong and complex magnetic field. Until the Voyager 2 flyby, Neptune was considered to have a rocky core surrounded by a liquid mantle. Spacecraft results, however, suggest that Neptune may have a low-density, liquid core. Laboratory experiments conducted by using a mixture of water, ammonia, and alcohol to simulate the interiors of both Neptune and Uranus show that, under the high pressures and temperatures considered to exist in the outer planets, a fluid core can be maintained and matches the postulated densities of both planets.

In contrast to the bland appearance of Uranus' atmosphere, Voyager images of Neptune show a wealth of cloud structures and motions (Fig. 11.1). Although the temperature in the upper atmosphere is a chilling 70 K, energy and heat released from the interior is apparently sufficient to cause convective cloud motions, in contrast to conditions on Uranus. The overall appearance of Neptune is blue, similar to Uranus and also a consequence of abundant methane in the atmosphere which absorbs red light.

Neptune's atmosphere travels with the greatest speed of any in the Solar System. While the planet rotates from west to east (like the Earth), winds at the equator travel in the opposite direction. Tracking of the clouds seen on sequential Voyager photographs shows that they travel more than 2000 km/hr at the equator.

Several distinctive cloud features were discovered by Voyager, as reported by Smith *et al.* (1989). The most pronounced, dubbed the Great Dark Spot (Figs. 11.1, 11.2), is found just below the equator at 20°S. This elongate feature is some 12,000 km wide (the size of the Earth) and rolls in a counter-clockwise direction, completing one rotation in about 16 days. Another dark spot, also found in the southern hemisphere, appears to be a smaller version of the Great Dark Spot. Both dark spots occur at a lower altitude within the clouds and hazes of the upper atmosphere.

Bright cirrus-like clouds are also visible on Neptune.

Figure 11.1 View of Neptune taken by the Voyager 2 spacecraft. The Great Dark Spot is the large prominent feature accompanied by bright white clouds. South of the Great Dark Spot is a smaller white cloud and a second dark spot (Voyager 2, JPL P-34648).

They appear to be dense, upward extensions of the methane clouds. Some are found in association with the Great Dark Spot; others occur as narrow belts in zones of wind shear in both the northern and southern hemispheres.

Neptune has a strong and complex magnetosphere and associated magnetotail. Measurements taken by Voyager 2 show that the magnetic field is offset from the center of Neptune and tilted about 47° from the rotational axis of the planet (Ness *et al.* 1989). In some ways, it is similar to the magnetic field observed at Uranus and, like that of Uranus, poses a puzzle. Although Neptune may also be undergoing a reversal of magnetic poles (as postulated for Uranus), for both planets to be experiencing such a reversal at the same time seems unlikely.

11.3 Rings

Rings (or more properly, segments of rings or ring-arcs) around Neptune were discovered independently by two teams of scientists in the mid-1980s. André Brahic and his colleagues and William Hubbard and Faith Villas noted occultations of stars while observing Neptune. The occultations were attributed to material in orbit around Neptune, similar to the discovery of the rings around Uranus. Subsequent pictures taken by Voyager showed several ring systems. Referred to as N63, N53 and N42 (in reference to the distances in thousands of kilometers from Neptune) these rings are all in prograde orbits confined to the equatorial plane of Neptune.

Both N63 and N53 are narrow rings (Fig. 11.3) composed of fine, dusty material, as evidenced by the way that they reflect sunlight. They appear to be somewhat

Figure 11.2 Sequential photographs of clouds near the Great Dark Spot, covering a period of about 36 hours, or two rotations of Neptune. The cirruslike clouds are thought to be composed of frozen methane (Voyager 2, JPL P-34622).

Figure 11.3 Voyager 2 image showing the two prominent rings, N 53 and N 63. Three "ring-arcs" (concentrations of dust particles) are seen in ring N 63. The direction of motion is clockwise (Voyager 2, JPL P-34712).

similar to the gamma ring of Uranus. N42 is a broad, diffuse ring which also seems to be composed of dust (Fig. 11.4), and is similar to Jupiter's ring and the G ring of Saturn. The two outer rings include clumps of dust that form rings, or ring-arcs. However, analysis of Voyager pictures shows that small amounts of dust fill in the space between the arcs to make complete rings.

A fourth ring mass forms a broad, diffuse sheet that extends throughout the inner Neptune system, as shown in Figure 11.4. In general, traveling from Neptune outward, one would find, first ring material, then small satellites and ring material, then the larger satellites. However, the entire mass of ring material around Neptune is some 10,000 times less than that of the uranian ring system and even less than that of Saturn's rings.

11.4 Satellites

Neptune has eight known satellites (Table 2.1), six of which were discovered from Voyager 2 data. The largest, Triton, is in the size class of Earth's Moon. The remainder range in diameter from 50 to 400 km. The four smallest satellites are in orbits within the ring system of Neptune, but do not appear to be ring shepherds. All are dark objects that travel in prograde orbits around Neptune. Data for these satellites are very limited, but from the manner in which they reflect light, the satellites probably have irregular shapes.

1989N1, the largest of the newly discovered satellites, orbits Neptune very rapidly just outside the ring system, taking only 1.1 Earth days to make one complete orbit.

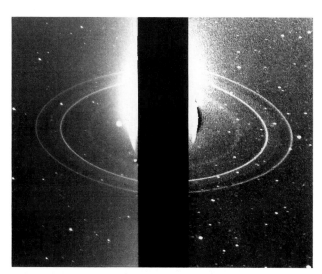

Figure 11.4 Image of the two main rings, N 63 and N 53, and the inner (faint) ring N 42. The fourth ring forms a faint band that extends from between the N 63 and N 53 rings inward toward Neptune. The bright spots are stars in the background (Voyager 2, JPL P-34726).

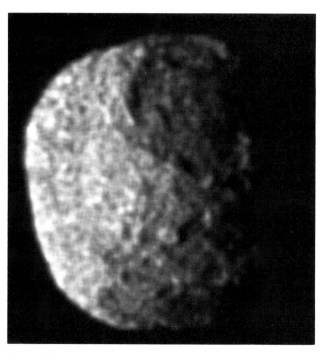

Figure 11.5 Image of 1989N1, the 400-km-diameter satellite of Neptune discovered during the Voyager 2 mission (Voyager 2, JPL P-34727).

Images (Fig. 11.5) show it to be irregular in shape and to have craters on the surface, some as large as 150 km across. In addition, an irregular trough-like depression can be seen on the surface. This feature could be a remnant of a crater, a structure formed by tectonic processes, or deformed crust resulting from a large impact.

Nereid is the outermost of Neptune's satellites and is about 340 km in diameter. It travels in a highly inclined, eccentric orbit and takes 359 days to complete one trip around Neptune. The Voyager spacecraft was unable to take high-resolution images of Nereid because it passed at too great a distance from the satellite; nor were any data obtained to allow estimates of its density. There are, however, some variations in brightness seen on the satellite, and these could be attributed either to an irregular shape or to variations in the composition of its surface.

11.5 Triton

Triton is a fascinating satellite that revealed many surprises to the Voyager cameras (Fig. 11.6). Not only does its surface show features unseen in the rest of the Solar System, but explosively erupting geysers were discovered. It is also one of the rare satellites to have an atmosphere, although it is very low in density, with an estimated surface pressure of less than 0.01 mb. Moreover, its orbital geometry is unique and poses some intriguing puzzles that have not yet been solved.

Figure 11.6 Photomosaic of Triton, showing the Neptune-facing hemisphere of the satellite. "Cantaloupe" terrain is clearly visible on the right side, as are smooth volcanic plains, dark plumes, and bright south polar deposits (left side) (Voyager 2, JPL P-34764).

Triton travels around Neptune in a circular, synchronous orbit but, unlike any other large satellite, it is in retrograde motion. In addition, careful analysis of Triton's motions shows that the satellite is in a decaying orbit, slowly being pulled toward Neptune. Its orbital geometry suggests that Triton was formed elsewhere in the Solar System, and was subsequently captured gravitationally by Neptune.

If Triton is a captured object, then in order to achieve its present orbit, it must have experienced extreme tidal stresses exerted by Neptune following its capture. These stresses undoubtedly would have generated large amounts of heat in the interior of the moon, perhaps leading to melting of the interior and compositional differentiation. At 2.1 g/cm^2 Triton has the highest density of any of the outer planet satellites. This suggests that it is composed of about one-third ice and two-thirds rocky materials. If Triton did experience heating and differentiation, then it would be expected to have a rocky core, surrounded by a mantle that could still be liquid, and a crust of ice.

Spectral reflectance measurements indicate that the surface of Triton is composed mostly of methane, nitrogen ices, and frost. Some of the areas are yellowish to peach-colored, which is thought to represent methane and nitrogen ices that have been converted to complex organic compounds from bombardment by cosmic rays and ultra-

violet radiation from the sun (Thompson and Sagan, 1990).

Two primary terrains are seen in Voyager images of Triton: an older rugged unit called *cantaloupe terrain* and younger plains considered to be of volcanic origin. Cantaloupe terrain is so named because of its resemblance to the skin of a melon and includes irregular pits, bumps, and linear veins (Fig. 11.7). The pits and bumps appear to represent extrusion of viscous ice onto the surface from the interior. Linear features include grooves and troughs, some of which are partly filled with ridges that also seem to be the result of the extrusion of ice along fracture systems. Some of the grooves appear to be grabens and exceed 1000 km in length. Impact craters pock-mark the cantaloupe terrain and provide some indication of the age of the surface, as discussed by Strom *et al.* (1990). Most of the craters are found on the leading hemisphere of Triton. The size and number of craters is about the same as for the basaltic maria on the Moon. Because of the lack of large craters and the absence of heavily cratered terrain, the age of Triton's surface is considered geologically young.

Relatively smooth plains cut across the cantaloupe terrain. This relationship and the near lack of impact craters show that the plains are younger than the cantaloupe terrain. The smooth plains are characterized by irregular-shaped depressions, some of which are 100 to 200 km across. These depressions appear to be similar to volcanic calderas. In some cases they may have been flooded with vast lakes, as shown in Figure 11.8. The calderas and volcanic plains are thought to have been formed by the eruption of liquid water, methane, and/or nitrogen onto

Figure 11.7 Detail of the "cantaloupe" terrain, showing linear grooves, irregular domes, and a general lack of impact craters (Voyager 2, JPL P-34694).

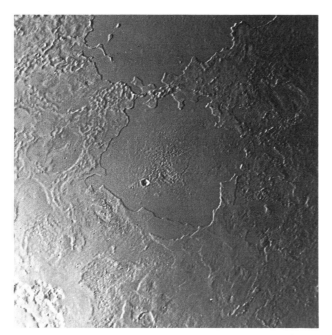

Figure 11.8 Smooth plains filling irregular calderalike depressions; area shown is about 500 km across (Voyager 2, JPL P-34692).

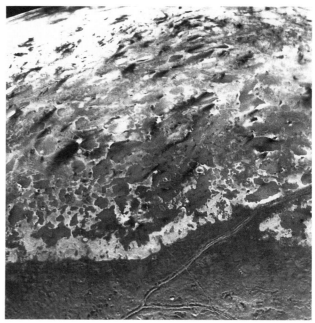

Figure 11.9 Image showing numerous dark streaks on Triton that result from geyserlike eruptions. The larger streaks exceed 100 km in length (Voyager 2, JPL P-34714).

the surface. Such a mixture would have a relatively low melting point and could have been produced in the interior of Triton.

Triton has a prominent south polar ice cap, probably composed mostly of nitrogen and methane ices which extend nearly to the equator. Moreover, the spin axis of Triton is inclined some 21° with respect to its orbital plane around Neptune. This means that like Earth and Mars, Triton also has seasons. With the changing seasons, the ice in one cap probably sublimates into the atmosphere and is redeposited at the other pole. At the time of the Voyager 2 flyby, in 1989, it was summer in the southern hemisphere and the south polar cap appears to have been sublimating gases into the atmosphere. As seen in Figure 11.6, the northern boundary of the ice cap is very irregular and follows local terrain. In addition, isolated patches of frost occur in depressions near the boundary.

Among the many discoveries made through the Voyager spacecraft, finding active eruptions on Io and Triton are among the highlights of the mission. During the initial analysis of Voyager data, two eruptive plumes were discovered on Triton (Smith *et al.* 1989). Later analysis of the data revealed two more active eruptions, as discussed by Soderblom and colleagues (1990). The eruptions consist of dark geysers as large as 1 kilometer across that extend vertically above the surface of Triton to an altitude of about 8 km. At this height, wind shear in the upper, thin atmosphere apparently catches the plume material

and carries it downwind some 100 km, leaving dark streaks on the surface (Fig. 11.9).

Several ideas have been proposed to explain Triton's eruptions (Brown *et al.* 1990; Kirk *et al.* 1990). Their location in the southern hemisphere in association with the ice cap may provide some clues to the process. First, it is known that nitrogen in its pure form is transparent. Nitrogen ice may serve as a kind of "solid state greenhouse", in which solar energy is trapped beneath the ice and raises the temperature at the base of the ice layer. A rise in temperature of only 4 K is adequate to vaporize the nitrogen ice. As more gas is released, the expanding vapors exert very high pressures beneath the ice cap. Eventually the gas ruptures the ice and is explosively released into the thin atmosphere above the surface. Dark material, perhaps silicates or ice particles darkened by ultraviolet radiation, is carried along with the expanding vapor into the atmosphere.

Regardless of how they are formed, the numerous dark streaks found on the surface suggest that geyser formation is common on Triton. Mapping the orientation of these and other features related to the atmosphere shows that most of the streaks are oriented toward the northeast and east. Analysis of these features and cloud patterns suggest that winds 1 to 3 km above the surface are predominantly eastward, whereas those at about 8 km blow toward the west.

12 Epilogue

The 1970s have been referred to as the "golden decade" of Solar System exploration (Murray, 1983). During this decade more than 34 successful missions were flown to the Moon and planets by the United States and the Soviet Union. As political interests shifted and economic forces prevailed, the beginning of the 1980s saw a marked decline in the number of missions. However, it must also be pointed out that both the quality and quantity of scientific data have increased significantly with time. Thus, the reduced number of missions was offset, at least partly, by better and more efficient spacecraft and instrumentation.

Some important "firsts" occurred in the 1980s, including an armada of five spacecraft sent to Halley's Comet. Especially important for planetary geology was the start of high-resolution mapping of Venus. In the late fall of 1983 the USSR put Veneras 15 and 16 into orbit around Venus. Carrying radar-imaging systems, these two spacecraft began systematic mapping of the northern hemisphere at spatial resolutions as good as 1–2 km from an orbital altitude of 1000–2000 km. The goal of mapping the entire area of Venus north of 35°N latitude was achieved during the nominal lifetime of the two spacecraft.

Exploration of Venus continued through the joint Soviet and French program, designated Vega 1 and 2. Designed principally for study of Halley's Comet, the "bus" dropped off a lander to obtain geochemical measurements of the venusian surface, and balloons were set adrift in the atmosphere. The Vega missions were very successful and contributed substantially to our knowledge not only of Halley's Comet, but of Venus as well.

Exploration of the outer Solar System continued with the journey of Voyager 2. In January 1986, this Mariner-class spacecraft observed the uranian system in much the same manner as did earlier flybys of Jupiter and Saturn and returned nearly 7000 images of Uranus and its rings and satellites (see Chapter 10). In August 1989 Voyager 2 encountered the Neptune system and returned thousands of images and other data (see Chapter 11).

With the explosion of the shuttle Challenger upon lift-off, early 1986 marked not only a tragic disaster for the manned near-Earth space program, but for planetary exploration as well. The use of expendable launch vehicles had been eliminated much earlier and all future planetary missions by the United States were to be launched from shuttle. The Galileo mission to Jupiter was to have

been launched in May 1986, but was delayed until 1990. Because the shuttle was unable to provide sufficient launch energy for a direct-to-Jupiter trajectory, the Galileo spacecraft was placed in orbit around the Sun for two years on a path that led to one flyby of Venus and two flybys of the Earth-Moon system. Each flyby enabled the spacecraft to gain energy through gravity assists so that it could be redirected toward Jupiter in late 1992. During its encounters with Venus and the Moon, new data were acquired. In addition, the trajectory carried Galileo into the asteroid belt, enabling the first-ever images of asteroids to be obtained (Belton *et al.* 1992; Figs. 12.1, 12.2). Unfortunately, the "high-gain" antenna on Galileo did not deploy as planned, severely crippling the ability of the spacecraft to return electronically the large amounts

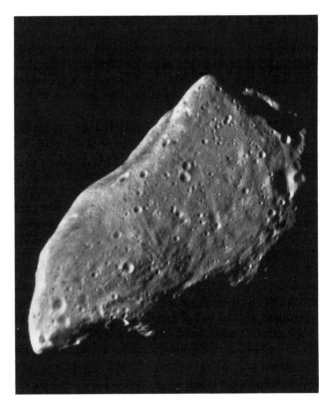

Figure 12.1 Galileo image of Gaspra, obtained 29 October 1991, from a distance of 1600 km. Gaspra is 19 × 12 × 11 km and orbits near the inner edge of the main asteroid belt. It is an S-type asteroid, thought to be composed of olivine and pyroxene with iron-nickel metals. The shape of Gaspra and its crater–size frequency suggest that this asteroid was formed by catastrophic collisional disruption of a larger parent body (JPL P-40449).

of data required by imaging systems. It is hoped that during the cruise to Jupiter (the spacecraft will arrive in late 1995), the antenna will deploy. Meanwhile, engineers are attempting to reprogram the spacecraft systems to enable more efficient transmission of data over the operational (but low-data rate) "low-gain" antenna. As discussed in Chapter 8, Galileo will release an entry probe into the atmosphere of Jupiter and orbit the jovian system for some 20 months. Imaging and other remote sensing data will enable a more complete understanding of this "miniature solar system."

In addition to the launch of Galileo, 1989 also saw the shuttle-launch of the Magellan radar orbiter to Venus. In less than two years of operation, more digital data were returned than from all previous planetary missions combined.

Earth's Moon continues to be a key object for planetary study. In a cooperative mission involving NASA and the Department of Defense, the Clementine mission involves the launch in early 1994 of an orbiter that will carry a variety of remote sensing instruments. After mapping the Moon, the spacecraft will be sent to an asteroid for detailed observations.

Mars is a high-priority objective for planetary exploration. Unfortunately, the Soviets' bad-luck attempts to send spacecraft to Mars continued with their Phobos mission launched in 1988. Two spacecraft carried an array of instruments, including landers designed for the martian satellite Phobos. An unfortunate error in commands caused loss of the first spacecraft less than two months after launch, and contact was lost with the second spacecraft shortly after it began orbital operations at Mars.

The future of the now-Russian planetary program is uncertain with the demise of the Soviet Union. Nonetheless, plans are moving forward for two Mars missions, dubbed Mars '94 and Mars '96 for the years of their anticipated launches. Mars '94 will involve an orbiter, two small landers, and two penetrators designed to slam into the planet and bury instruments beneath the surface. Mars '96 would be similar, but with the addition of two French balloons which would be released in the atmosphere to drift for some 10 days. During night-time the balloons would settle to the surface to take measurements; with solar heating the next day, the balloons would rise above the surface and be carried downwind to a new location and the process would be repeated. Plans also call for a small ($\sim$1.5 m) roving vehicle to be released from each balloon during the first touchdown.

The Mars '94 and '96 missions are truly international, with instruments being provided by many countries and with the participation of scientists from around the world, including the United States.

The Mars Observer mission was launched successfully by the United States in September 1992. After a cruise of nearly one year, the spacecraft arrived at Mars in August, 1993. During the operations to place Mars Observer into orbit, communication was lost and the mission failed. Although plans to refly the mission are not firm, the scientific community is unanimous in the opinion that the measurements involving global geochemical and geophysical surveys, imaging, and atmospheric measurements are of the highest priority.

After analyses of Mars-94/96 data and measurements from a Mars Observer-like mission, most of the remaining first-order questions for Mars will require measurements from landed spacecraft. For example, little is known about the interior characteristics (core, mantle, etc.) of the planet. Consequently, networks of multiple small stations are being considered for deployment by the United States and the European Space Agency (ESA). The US mission, termed MESUR (Mars Environmental Survey), and the ESA MARSNET mission could place as many as 20 stations on Mars for operation into the next century. All would include seismometers, weather stations, imaging systems, and instruments designed to measure rock and soil compositions.

Plans for continued outer Solar System exploration include the Cassini mission, approved as an orbiter of the Saturn system, and discussions of a flyby mission for Pluto and its moon, Charon. The Cassini mission is a joint venture with ESA. It would include a long-term orbiter carrying remote sensing instruments, a radar-imaging system to observe the cloud-shrouded saturnian satellite, Titan, and a probe to examine the atmosphere of Saturn.

After their successful flight to Halley's Comet, the Japanese plan to continue planetary exploration with missions to the Moon and Mars before the end of the century. Their lunar mission, to be launched in 1997, would involve a polar orbiter and surface penetrators that would implant seismometers, heat-flow instruments, and devices to measure the chemistry of the surface of the Moon. The Mars mission is designed to study the interaction of the solar wind with the atmosphere, assess the ionosphere, and obtain global views of atmospheric clouds.

The United States also has tentative plans to return to the Moon with polar orbiting spacecraft and possible automated landers. These missions may be linked to a program leading to the return of humans in direct planetary exploration.

The US, the Russians, ESA and the Japanese have all considered more ambitious plans for future missions, either through separate efforts or joint programs. These include samples returned from Mars, asteroids and other solid-surface objects, and planetary outposts on the Moon and Mars staffed by scientists, engineers and technicians.

The last decade has seen an enormous change in the manner in which planetary research is conducted. Advances in small computer workstations, coupled with increasingly widespread distribution of digital data on CD

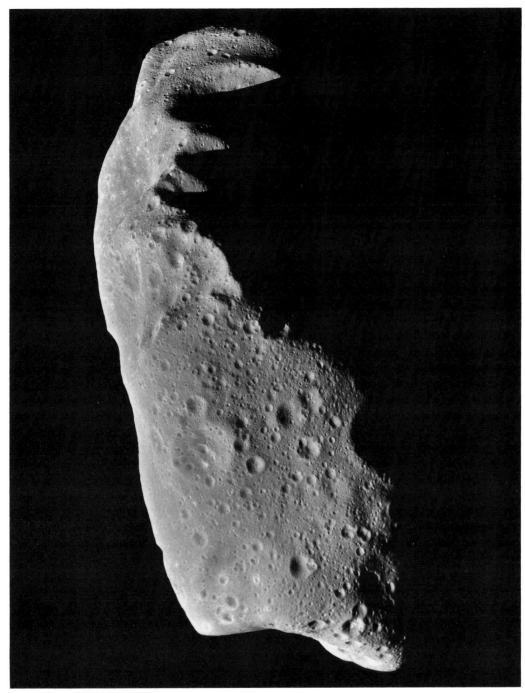

Figure 12.2 This mosaic of five images was taken of the asteroid Ida by Galileo in August, 1993. Ida is an S-type asteroid and is about 55 km long. Differences in albedo on crater floors, irregular shaped crater rims, and features that may be blocks on the surface all suggest that the asteroid has a regolith or fragmental surface layer (JPL P-42964).

ROMs, mean that many more scientists can carry out first-order research using spacecraft data. The next decade promises still greater advances through technology. Development of sophisticated robots, microrovers less than 1 m in size, and the evolution of "virtual reality" will change the approaches used in planetary exploration.

While the list of planned, potential, and postulated missions for planetary exploration is impressive, we must recognize that selection and final approval are in many ways as hazardous as space travel itself and are frequently driven by political and financial considerations. Nonetheless, whichever missions are ultimately flown, geoscience figures prominently in the analyses of the data for the solid-surface objects and we look forward to an exciting era of continued exploration of the Solar System.

References

Arvidson, R. E., Goettel, K. A. and Hohenberg, C. M. (1980). A post-Viking view of Martian geologic evolution. *Rev. Geophys. Space Phys.* **18**, 565–603.

Bagnold, R. A. (1941). *The physics of blown sand and desert dunes*. London, Methuen.

Baker, V. R. (1982). *The channels of Mars*. University of Texas Press, Austin.

Baker, V. R. (1984). Planetary geomorphology. *J. Geol. Educ.* **32**, 236–46.

Baker, V. R. and Kochel, R. C. (1979). Martian channel morphology: Maja and Kasei Valles. *J. Geophys. Res.* **84**, 7961–83.

Baker, V. R. and Nummedal, D. (eds) (1978). The *channeled scabland: a guide to the geomorphology of the Columbia Basin, Washington*. Washington, DC, NASA, Planet. Geol. Prog.

Baker, V. R., Komatsu, G., Parker, T. J., Gulick, V. C., Kargel, J. S. and Lewis, J. S. (1992). Channels and valleys on Venus: Preliminary analysis of Magellan data. *J. Geophys. Res.* **97**, 13 421–44.

Baldwin, R. B. (1949). *The face of the Moon*. University of Chicago Press, Chicago.

Baldwin, R. B. (1963). *The measure of the Moon*. University of Chicago Press, Chicago.

Barsukov, V. L., Volkov, V. P., and Khodakovsky, I. L. (1982). The crust of Venus: theoretical models of chemical and mineral composition. *J. Geophys. Res.* **87**, suppl. A3–A9.

Barsukov, V. L., Basilevsky, A. T., Volkov, V. P., and Zharkov, V. N. (eds) (1992). *Venus Geology, Geochemistry, and Geophysics*. University of Arizona Press, Tucson.

Basilevsky, A. T. and Head III, J. W. (1988). The geology of Venus. *In* Wetherill, G. W., A. L. Albee, and F. G. Stehli (eds) *Annual Review of Earth and Planetary Sciences* **16**, 295–317.

Batson, R. M. (1981). Status and future of extraterrestrial mapping programs. *NASA CR 3390*.

Beatty, J. K. and Chaikin, A. (eds) (1990). *The new Solar System*. Cambridge University Press, Cambridge.

Belton, M. J. S., Hunten, D. M. and Goody, R. M. (1968). Quantitative spectroscopy of Venus in the region 8,000–11,000 Å. In *The atmospheres of Venus and Mars*, J. C. Brandt and M. B. McElroy (eds), 69–98. Gordon and Breach, New York.

Belton, M. J. S. and the Galileo imaging team (1992). Galileo encounter with 951 Gaspra: First pictures of an asteroid. *Science* **257**, 1647–52.

Bergstralh, J. T., Miner, E. D., and Matthews, M. S. (eds) (1991). *Uranus*. University of Arizona Press, Tucson.

Bindschadler, D. L., deCharon, A., Beratan, K. K., Smrekar, S. E. and Head, J. W. (1992*a*). Magellan observations of Alpha Regio: Implications for formation of complex ridged terrains on Venus. *J. Geophys. Res.* **97**, 13 563–77.

Bindschadler, D. L., Schubert, G., and Kaula, W. M. (1992*b*). Coldspots and hotspots: Global tectonics and mantle dynamics of Venus. *J. Geophys. Res.* **97**, 13 495–532.

Bloom, A. L. (1978). *Geomorphology*. Prentice-Hall, Englewood Cliffs.

Blunck, J. (1977). *Mars and its satellites: a detailed commentary on the nomenclature*. Exposition Press, Hicksville.

Brown, R. H., Kirk, R. L., Johnson, T. V. and Soderblom, L. A. (1990). Energy sources for Triton's geyser-like plumes. *Science* **250**, 431–5.

Burns, J. A. (ed) (1977). *Planetary satellites*. University of Arizona Press, Tucson.

Burns, J. A. and Matthews, M. S. (eds) (1986). *Satellites*. University of Arizona Press, Tucson.

BVSP (Basaltic Volcanism Study Project) (1981). *Balsatic volcanism on the terrestrial planets*. Pergamon Press, New York.

Calcogno, L., Foti, G., Torrisi, L. and Strazzulla, G. (1985). Fluffy layers obtained by ion bombardment of frozen methane: Experiments and applications to saturnian and uranian satellites. *Icarus* **63**, 31–8.

Carr, M. H. (1973). Volcanism on Mars. *J. Geophys. Res.* **78**, 4049–62.

Carr, M. H. (1979). Formation of martian flood features by release of water from confined aquifers. *J. Geophys. Res.* **84**, 2995–3007.

Carr, M. H. (1980). The morphology of the martian surface. *Space Sci. Rev.* **25**, 231–84.

Carr, M. H. (1981). *The surface of Mars*. Yale University Press, New Haven.

Carr, M. H. (1983). The geology of the terrestrial planets. *U.S. National Report, 1979–1982*, Am. Geophys. Union, 160–72.

Carr, M. H. and Schaber, G. G. (1977). Martian permafrost features. *J. Geophys. Res.* **82**, 4039–54.

Carr, M. H., Crumpler, L. S., Cutts, J. A., Greeley, R., Guest, J. E. and Masursky, H. (1977). Martian impact craters and emplacement of ejecta by surface flow. *J. Geophys. Res.* **82**, 4055–65.

Carr, M. H., Masursky, H., Strom, R. G. and Terrile, R. J. (1979). Volcanic features of Io. *Nature* **280**, 729–33.

Cassen, P., Peale, S. J. and Reynolds, R. T. (1980). On the comparative evolution of Ganymede and Callisto. *Icarus* **41**, 232–9.

Cattermole, P. (1992). *Mars: The Story of the Red Planet.* Chapman and Hall, London.

Chapman, C. R. and Jones, K. L. (1977). Cratering and obliteration history of Mars. *Ann. Rev. Earth Planet. Sci.* **5**, 515–40.

Cintala, M. J., Wood, C. A. and Head, J. W. (1977). The effects of target characteristics on fresh crater morphology: preliminary results for the Moon and Mercury. *Proc. Lunar Sci. Conf. 8th*, 3409–25.

Classen, J. (1977). Catalogue of 230 certain, probable, possible, and doubtful impact structures. *Meteoritics* **12**, 61–78.

CLHC (1979). *Proceedings of the conference on the lunar highlands.* Lunar Planetary Institute, Houston.

Clifford, S. M. and Hillel, D. (1983). The stability of ground ice in the equatorial region of Mars. *J. Geophys. Res.* **88**, 2456–74.

Clow, G. D. and Carr, M. H. (1980). Stability of sulfur slopes on Io. *Icarus* **44**, 268–79.

CMRB (1980). *Proceedings of the conference on multi-ring basins.* Lunar Planetary Institute, Houston.

Colin, L. (1980). The Pioneer Venus program. *J. Geophys. Res.* **85**, 7575–98.

Colwell, R. N. (ed.) (1983). *Manual of remote sensing,* vols. 1 and 2, 2nd edn. American Society of Photogrammetry, Falls Church.

Condit, C. D. and Chavez, P. S. (1979). Basic concepts of computerized digital image processing for geologists. *U.S. Geol. Surv. Bull.*, No. 1462.

Cook, A. F. and Franklin, F. A. (1970). An explanation of the light curve of Iapetus. *Icarus* **13**, 282–91.

Cotton, C. A. (1952). *Volcanoes as landscape forms.* Whitcomb and Tombs (reprinted 1969, Hafner, New York).

Croft, S. K. (1983). A proposed origin for palimpsests and anomalous pit craters on Ganymede and Callisto. *J. Geophys. Res.* **88**, suppl. 71–89.

Cruikshank, D. P., Degewij, J. and Zellner, B. H. (1982). The outer satellites of Jupiter. In *Satellites of Jupiter,* D. Morrison (ed.), 129–46. University of Arizona Press, Tucson.

Cruikshank, D. P., Bell, J. F., Gaffey, M. J., Brown, R. H., Howell, R., Beerman, C. and Rongstad, M. (1983). The dark side of Iapetus. *Icarus* **53**, 90–104.

Crumpler, L. S. and Aubele, J. C. (1978). Structural evolution of Arsia Mons, Pavonis Mons, and Ascraeus Mons: Tharsis region of Mars. *Icarus* **34**, 496–511.

Cutts, J. A. and Blasius, K. R. (1981). Origin of martian outflow channels: the eolian hypothesis. *J. Geophys. Res.* **86**, 5075–102.

Cutts, J. A. and Lewis, B. H. (1982). Models of climate cycles recorded in martian polar layered deposits. *Icarus* **50**, 216–44.

Cutts, J. A., Thompson, T. W. and Lewis, B. H. (1981). Origin of bright ring-shaped craters in radar images of Venus. *Icarus* **48**, 428–52.

Dalrymple, G. B., Silver, E. A. and Jackson, E. D. (1973). Origin of the Hawaiian Islands. *Am. Scientist* **61**, 294–308.

Davies, M. E. and Batson, R. M. (1975). Surface coordinates and cartography of Mercury. *J. Geophys. Res.* **80**, 2417–30.

Davies, M. E. and Murray, B. C. (1971). *The view from space: photographic exploration of the planets.* Columbia University Press, New York.

Davies, M. E., Dwornik, S. E., Gault, D. E. and Strom, R. G. (1978). *Atlas of Mercury.* NASA SP-423.

Davis, W. M. (1926). Biographical memoir of Grove Karl Gilbert, 1843–1918. *Memoir Nat. Acad. Sci.* **21**, 5th Mem., 303.

DeHon, R. A. (1978). In search of ancient astroblemes: Mercury. In *Rep. Planet. Geol. Prog.—1977–1978,* NASA TM-79729, 150–2.

DeHon, R. A. (1979). Thickness of the western Mare Basalts. *Proc. Lunar Planet. Sci. Conf. 10th*, 2935–55.

Dence, M. R. (1972). The nature and significance of terrestrial impact structures. *Proc. 24th Int. Geol. Congr.*, Montreal, Section 15, 77–89.

Dence, M. R. (1976). Notes toward an impact model for the Imbrium basin. *Interdisc. Stud. Imbrium Consort.* **1**, 147–55.

Donahue, T. M., Hoffman, J. H., Hodges, Jr., R. R. and Watson, A. J. (1982). Venus was wet: a measurement of the ratio of deuterium to hydrogen. *Science* **216**, 630–3.

Dzurisin, D. (1978). The tectonic and volcanic history of Mercury as inferred from studies of scarps, ridges, troughs, and other lineaments. *J. Geophys. Res.* **83**, 4883–906.

El-Baz, F. (1979). Scientific exploration of the Moon. *Interdisc. Sci. Rev.* **4**, 239–61.

Ellsworth, K. and Schubert, G. (1983). Saturn's icy satellites: thermal and structural models. *Icarus* **54**, 490–510.

Eppler, D. and Heiken, G. (1975). Lunar crater chains of non-impact origin. *Proc. Lunar Planet. Sci. Conf. 6th*, 2571–83.

Fanale, F. P. (1976). Martian volatiles: their degassing history and geochemical fate. *Icarus* **28**, 179–202.

Fanale, F. P., Brown, R. H., Cruikshank, D. P. and Clark, R. N. (1979). Significance of absorption features in Io's IR reflectance spectrum. *Nature* **280**, 761–3.

Farmer, C. B., Davies, D. W., Holland, A. L., La Porte, D. D. and Doms, P. E. (1977). Mars: water vapor observations from the Viking orbiters. *J. Geophys. Res.* **82**, 4225–48.

Fielder, G. (1961). *Structure of the Moon's surface.* Pergamon, London.

Fielder, G. (1965). *Lunar geology.* Dufour, Chester Springs.

Fimmel, R. O., Swindell, W. and Burgess, E. (1977). *Pioneer odyssey.* NASA SP-396.

Fink, J. H., Park, S. O. and Greeley, R. (1983). Cooling and deformation of sulfur flows. *Icarus* **56**, 38–50.

Finnerty, A. A., Ransford, G. A., Pieri, D. C. and Collerson, K. D. (1980). Is Europa surface cracking due to thermal evolution? *Nature* **289**, 24–7.

Francis, P. W. and Wood, C. A. (1982). Absence of silicic volcanism on Mars: implications for crustal composition and volatile abundance. *J. Geophys. Res.* **87**, 9881–9.

French, B. M. and Short, N. M. (eds) (1968). *Shock metamorphism of natural materials.* Mono Book, Baltimore.

Frey, H. and Lowry, B. L. (1979). Large impact basins on Mercury and relative crater production rates. *Proc. Lunar Planet. Sci. Conf. 10th,* 2669–87.

Frey, H., Lowry, B. L. and Chase, S. A. (1979). Pseudocraters on Mars. *J. Geophys. Res.* **84**, 8075–86.

Gault, D. E. (1970). Saturation and equilibrium conditions for impact cratering on the lunar surface: criteria and implications. *Radio Sci.* **5**, 273–91.

Gault, D. E. (1974). Impact craters. In *Primer in lunar geology,* R. Greeley and P. E. Schultz (eds), 137–75. NASA TM X 62359.

Gault, D. E. and Greeley, R. (1978). Exploratory experiments of impact craters formed in viscous-liquid targets: analogs for martian rampart craters? *Icarus* **34**, 486–95.

Gault, D. E. and Wedekind, J. A. (1978). Experimental studies of oblique impact. *Proc. Lunar Planet. Sci. Conf. 9th,* 3843–75.

Gault, D. E., Quaide, W. L. and Oberbeck, V. R. (1968). Impact cratering mechanics and structures. In *Shock metamorphism of natural materials,* B. M. French and N. M. Short (eds.), 87–99. Mono Book, Baltimore.

Gault, D. E., Guest, J. E., Murray, J. B., Dzurisin, D. and Malin, M. C. (1975). Some comparisons of impact craters on Mercury and the Moon. *J. Geophys. Res.* **80**, 2444–60.

Gault, D. E., Burns, J. A., Cassen, P. and Strom, R. G. (1977). Mercury. *Ann. Rev. Astron. Astrophys.* **15**, 97–126.

Gehrels, T. and Matthews, M. S. (eds.) (1984). *Saturn.* University of Arizona Press, Tucson.

Gilbert, G. K. (1893). The Moon's face—a study of the origin and its features. *Phil Soc. Washington Bull.* **12**, 241–92.

Glass, B. P. (1982). *Introduction to planetary geology.* Cambridge University Press, Cambridge.

Glen, J. W. (1974). *The physics of ice.* Cold Regions Research and Engineering Laboratory, US Army, Monograph II-C2a.

Golombek, M. P. (1979). Structural analysis of lunar grabens and the shallow crustal structure of the Moon. *J. Geophys. Res.* **84**, 4657–66.

Golombek, M. P. (1982). Constraints on the expansion of Ganymede and the thickness of the lithosphere. *J. Geophys. Res.* **87**, suppl., 77–83.

Greeley, R. (1971). Lunar Hadley Rille: considerations of its origin. *Science* **172**, 722–5.

Greeley, R. (1976). Modes of emplacement of basalt terrains and an analysis of mare volcanism in the Orientale Basin. *Proc. Lunar Sci. Conf. 7th,* 2747–59.

Greeley, R. and Batson, R. M. (1990). *Planetary Mapping.* Cambridge University Press, Cambridge.

Greeley, R. and Carr, M. H. (eds) (1976). *A geological basis for the exploration of the planets,* NASA SP-417.

Greeley, R. and Gault, D. E. (1979). Endogenic craters on basaltic lava flows: size frequency distributions. *Proc. Lunar Planet. Sci. Conf. 10th,* 2919–33.

Greeley, R. and Iversen, J. D. (1985). *Wind as a geological process.* Cambridge University Press, Planetary Science Series, Cambridge.

Greeley, R. and Spudis, P. (1978a). Mare volcanism in the Herigonius region of the Moon. *Proc. Lunar Planet. Sci. Conf. 9th,* 3333–49.

Greeley, R. and Spudis, P. D. (1978b). Volcanism in the cratered terrain hemisphere of Mars. *Geophys. Res. Lett,* **5**, 453–5.

Greeley, R. and Spudis, P. D. (1981). Volcanism on Mars. *Rev. Geophys. Space Phys.* **19**, 13–41.

Greeley, R. and Theilig, E. (1978). Small volcanic constructs in the Chryse Planitia region of Mars. In *Rep. Planet. Geol. Prog.—1977–1978,* NASA TM-79729, 202.

Greeley, R., Theilig, E., Guest, J. E., Carr, M. H., Masursky, H. and Cutts, J. A. (1977). Geology of Chryse Planitia. *J. Geophys. Res.* **82**, 4093–109.

Greeley, R., Arvidson, R. E., Elachi, C., Geringer, M. A., Plaut, J. J., Saunders, R. S., Schubert, G., Stofan, E. R., Thouvenot, E. J. P., Wall, S. D. and Weitz, C. M. (1992). Aeolian features on Venus: Preliminary Magellan results. *J. Geophys. Res.* **97**, 13 319–45.

Greeley, R., Fink, J., Gault, D. E., Snyder, D. B., Guest, J. E. and Schultz, P. H. (1980). Impact cratering in viscous targets: laboratory experiments. *Proc. Lunar Planet. Sci. Conf. 11th,* 2075–97.

Greeley, R., Fink, J. H., Gault, D. E. and Guest, J. E. (1982). Experimental simulation of impact cratering on icy satellites. In *Satellites of Jupiter,* D. Morrison (ed.), 340–78. University of Arizona Press, Tucson.

Green, J. (1965). Hookes and spurs in selenology. In *Geological problems in lunar research, Ann. New York Acad. Sci.* **123**, 373–402.

Green, J. and Short, N. M. (1971). *Volcanic landforms*

and surface features: a photographic atlas and glossary. Springer, New York.

Greenberg, R. and Brahic, A. (eds) (1984). *Planetary Rings.* University of Arizona Press, Tucson.

Grieve, R. A. F. and Head, J. W. (1982). The impact cratering process on Venus. *Lunar Planet. Sci.* **13**, 285–6.

Guest, J. E. (1971). Centers of igneous activity in maria. In *Geology and physics of the Moon,* G. Fielder (ed.), 41–53. Elsevier, Amsterdam.

Guest, J. E. and Gault, D. E. (1976). Crater populations in the early history of Mercury. *Geophys. Res. Lett.* **3**, 121–3.

Guest, J. E. and Greeley, R. (1977). *Geology on the Moon.* Russak, New York.

Guest, J. E. and Murray, J. B. (1976). Volcanic features of the nearside equatorial lunar maria. *J. Geol. Soc. Lond.* **132**, 251–8.

Guest, J. E. and O'Donnell, W. P. (1977). Surface history of Mercury: a review. *Vistas Astron.* **20**, 273–300.

Guest, J. E., Butterworth, P. S. and Greeley, R. (1977). Geological observations in the Cydonia region of Mars from Viking. *J. Geophys. Res.* **82**, 4111–20.

Guest, J., with Butterworth, P., Murray, J. and O'Donnell, W. (1979). *Planetary geology.* Halsted, New York.

Guest, J. E., Bulmer, M. H., Aubele, J., Beratan, K., Greeley, R., Head, J. W., Michaels, G., Weitz, C. and Wiles, C. (1993). Small volcanic edifices and volcanism in the plains of Venus. *J. Geophys. Res.* (to appear).

Hale, W. S. and Grieve, R. A. F. (1982). Volumetric analysis of complex lunar craters: implications for basin ring formation. *J. Geophys. Res.* **87**, A65–A76.

Hale, W., Head, J. W. and Parmentier, E. M. (1980). Origin of the Valhalla ring structure: alternative models. In *Conference on multi-ring basins,* 30–2. Lunar Planetary Institute, Houston.

Hall, J. L., Solomon, S. C. and Head, J. W. (1981). Lunar floor-fractured craters: evidence for viscous relaxation of crater topography. *J. Geophys. Res.* **86**, 9537–52.

Hamblin, W. K. and Christiansen, E. H. (1990). *Exploring the Planets.* Macmillan Publishing Company, New York.

Hanel, R., Conrath, B., Flasar, F. M., Kunde, V., Maguire, W., Pearl, J., Pirraglia, J., Samuelson, R., Herath, L., Allison, M., Cruikshank, D., Gautier, D., Gierasch, P., Horn, L., Koppany, R., and Ponnamperuma, C. (1981). Infrared observations of the saturnian system from Voyager 1. *Science* **212**, 192–200.

Harris, S. A. (1977). The aureole of Olympus Mons, Mars. *J. Geophys. Res.* **82**, 3099–107.

Hartmann, W. K. (1983). *Moon and planets.* Wadsworth, Belmont.

Hartmann, W. K., Strom, R. G., Weidenschilling, S. J., Blasius, K. R., Woronow, A., Dence, M. R., Grieve, R.A.F., Diaz, J., Chapman, C. R., Shoemaker, E. M.

and Jones, K. L. (1981). Chronology of planetary volcanism by comparative studies of planetary cratering. In *Basaltic volcanism on the terrestrial planets.* Basaltic Volcanism Study Project, 1048–127. Pergamon Press, New York.

Hawke, B. R. and Head, J. W. (1978). Lunar KREEP volcanism: geological evidence for history and mode of emplacement. *Proc. Lunar Planet. Sci. Conf. 9th,* 3285–309.

Head, J. W. (1976). Lunar volcanism in space and time. *Rev. Geophys. Space Phys.* **14**, 265–300.

Head, J. W. and Gifford, A. (1980). Lunar mare domes: classification and modes of origin. *Moon Planets* **22**, 235–58.

Head, J. W. and Solomon, S. C. (1981). Tectonic evolution of the terrestrial planets. *Science* **213**, 62–76.

Head, J. W. III, Bryan, W. B., Greeley, R., Guest, J. E., Schultz, P. H., Sparks, R. S. J., Walker, G. P. L., Whitford-Stark, J. L., Wood, C. A. and Carr, M. H. (1981). Distribution and morphology of basaltic deposits on planets. In *Basaltic volcanism on the terrestrial planets,* Basaltic Volcanism Study Project, 701–800. Pergamon Press, New York.

Head, J. W., Campbell, D. B., Elachi, C., Guest, J. E., McKenzie, D. P., Saunders, R. S., Schaber, G. G. and Schubert, G. (1991). Venus volcanism: initial analysis from Magellan data. *Science* **252**, 276–88.

Head, J. W., Crumpler, L. S., Aubele, J. C., Guest, J. E. and Saunders, R. S. (1992). Venus Volcanism: classification of volcanic features and structures, associations, and global distribution from Magellan data. *J. Geophys. Res.* **97**, 13 153–97.

Heiken, G. H., Vaniman, D. T. and French, B. M. (1991). *Lunar Source Book.* Cambridge University Press, Cambridge.

Helfenstein, P. and Parmentier, E. M. (1980). Fractures on Europa: possible response of an ice crust to tidal deformation. *Proc. Lunar Planet. Sci. Conf. 11th,* 1987–98.

Hodges, C. A. (1973). Mare ridges and lava lakes. *Apollo 17 Prelim. Sci. Rep.,* NASA SP-330, 31. 12–31.21.

Hodges, C. A. and Moore, H. J. (1979). The subglacial birth of Olympus Mons and its aureoles. *J. Geophys. Res.* **84**, 8061–74.

Hodges, C. A. and Wilhelms, D. E. (1978). Formation of lunar basin rings. *Icarus* **34**, 294–323.

Horner, V. M. and Greeley, R. (1982). Pedestal craters on Ganymede. *Icarus* **51**, 549–62.

Hörz, F. (1978). How thick are lunar mare basalts? *Proc. Lunar Planet. Sci. Conf. 9th,* 3311–31.

Howard, A. D. (1967). Drainage analysis in geological interpretation: a summation. *Am. Assoc. Petrol. Geol. Bull.* **51**, 2246–59.

Howard, K. A. (1973). Avalanche mode of motion: implications from lunar examples. *Science* **180**, 1052–5.

Howard, K. A., Head, J. W. and Swann, G. A. (1972).

Geology of Hadley Rille. *Proc. Lunar Sci. Conf. 3rd,* 1–14.

Howard, K. A., Wilhelms, D. E. and Scott, D. H. (1974). Lunar basin formation and highland stratigraphy. *Rev. Geophys. Space Phys.* **12**, 309–27.

Hunt, G. (ed.) (1982). *Uranus and the outer planets.* Cambridge University Press, Cambridge.

Hunten, D. M., Colin, L., Donahue, T. M. and Moroz, V. I. (eds) (1983). *Venus.* University of Arizona Press, Tucson.

Inge, J. L. and Batson, R. M. (1992). *Indexes of maps of the planets and satellites.* National Aeronautics and Space Administration Tech. Memo. 4395.

Inge, J. L. and Bridges, P. M. (1976). Applied photo interpretation of airbrush cartography. *Photogram. Engng Remote Sens.* **42**, 749–60.

Johnson, T. V. (1978). The Galilean satellites of Jupiter: four worlds. *Ann. Rev. Earth Planet. Sci.* **6**, 93–125.

Johnson, T. V. and McGetchin, T. R. (1973). Topography on satellite surfaces and the shape of asteroids. *Icarus* **18**, 612–20.

Johnson, T. V. and Soderblom, L. A. (1982). Volcanic eruptions on Io: implications for surface evolution and mass loss. In *Satellites of Jupiter,* D. Morrison (ed.), 634–46. University of Arizona Press, Tucson.

Johnson, T. V., Cook II, A. F., Sagan, C. and Soderblom, L. A. (1979). Volcanic resurfacing rates and implications for volatiles on Io. *Nature* **280**, 746–50.

Kieffer, H. H. and Palluconi, F. D. (1979). *The climate of the martian polar cap.* NASA CP-2072, 45–6.

Kieffer, S. W. (1982). Dynamics and thermodynamics of volcanic eruptions: implications for the plumes on Io. In *Satellites of Jupiter,* D. Morrison (ed.), 647–723. University of Arizona Press, Tucson.

King, E. A. (1976). *Space geology.* John Wiley, New York.

King, J. S. and Riehle, J. R. (1974). A proposed origin of the Olympus Mons escarpment. *Icarus* **23**, 300–17.

King, L. C. (1967). *The morphology of the Earth.* Oliver and Boyd, Edinburgh.

Kirk, R. L., Brown, R. H. and Soderblom, L. A. (1990). Subsurface energy storage and transport for solar-powered geysers on Triton. *Science* **250**, 424–29.

Klein, H. P. (1979). The search for life on Mars. *Rev. Geophys. Space Phys.* **17**, 1655–62.

Kozai, Y. (1976). Masses of satellites and oblateness parameters of Saturn. *Publn Astron. Soc. Jpn* **28**, 675–91.

Kuiper, G. P. (1962). Infrared spectra of stars and planets. I. Photometry of the infrared spectrum of Venus, 1–2.5 microns. *Comm. Lunar Planet. Lab.* **1**, 83–117.

Lanzerotti, L. J., Brown, W. L., Poate, J. M. and Augustyniak, W. M. (1978). On the contribution of water products from Galilean satellites to the jovian magnetosphere. *Geophys. Res. Lett.* **5**, 155–8.

Leake, M. (1981). The intercrater plains of Mercury and the Moon: their nature, origin, and role in terrestrial planet evolution. Ph.D. dissertation, Univ. of Arizona, Tucson, AZ, also in *Adv. Planet. Geol.,* NASA TM 84894, Sept. 1982, 1–535.

Leopold, L. B., Wolman, M. G. and Miller, J. P. (1964). *Fluvial processes in geomorphology.* W. H. Freeman, San Francisco.

Lewis, J. S. (1971). Satellites of the outer planets: their physical and chemical nature. *Icarus* **15**, 174–85.

Lopes, R. M. C., Guest, J. E. and Wilson, C. J. (1980). Origin of the Olympus Mons aureole and perimeter scarp. *Moon Planets* **22**, 221–34.

Lopes, R., Guest, J. E., Hiller, K. and Neukum, G. (1982). Further evidence for a mass movement origin of the Olympus Mons aureole. *J. Geophys. Res.* **87**, 9917–28.

Lowman, P. D. Jr. (1981). *A global tectonic activity map with orbital photographic supplement.* NASA TM-82073.

Lucchitta, B. K. (1976). Mare ridges and related highland scarps—result of vertical tectonism? *Proc. Lunar Planet. Sci. Conf. 7th,* 2761–82.

Lucchitta, B. K. (1979). Landslides in Valles Marineris, Mars. *J. Geophys. Res.* **84**, 8097–113.

Lucchitta, B. K. (1981). Mars and Earth: comparison of cold-climate features. *Icarus* **45**, 264–303.

Lucchitta, B. K. (1982). Ice sculpture in the martian outflow channels. *J. Geophys. Res.* **87**, 9951–73.

Lucchitta, B. K. and Ferguson, H. M. (1983). Chryse Basin channels: low-gradients and ponded flows. *J. Geophys. Res.* **88**, suppl., A553–68.

Lucchitta, B. K. and Klockenbrink, J. L. (1981). Ridges and scarps in the equatorial belt of Mars. *Moon Planets* **24**, 415–29.

Lucchitta, B. K. and Soderblom, L. A. (1982). The geology of Europa. In *Satellites of Jupiter,* D. Morrison (ed.), 521–55. University of Arizona Press, Tucson.

Lucchitta, B. K. and Watkins, J. A. (1978). Age of graben systems on the Moon. *Proc. Lunar Planet. Sci. Conf. 9th,* 3459–72.

Lucchitta, B. K., Soderblom, L. A., and Ferguson, H. M. (1981). Structures on Europa. *Proc. Lunar Planet. Sci.* **12B**, 1555–67.

Lunine, J. I., Neugebauer, G. and Jakoski, B. M. (1982). Infrared observations of Phobos and Deimos from Viking. *J. Geophys. Res.* **87**, 10297–305.

Lupo, M. J. and Lewis, J. S. (1980). Mass-radius relationships and constraints on the composition of Pluto, II. *Icarus* **44**, 41–2.

McCauley, J. F. (1977). Orientale and Caloris. *Phys. Earth Planet. Inter.* **15**, 220–50.

McCauley, J. F. (1978). *Geologic map of the Coprates*

quadrangle of Mars. US Geol Survey, Misc. Inv. Map I-897.

McCauley, J. F., Smith, B. A. and Soderblom, L. A. (1979). Erosional scarps on Io. *Nature* **280**, 736–8.

McCauley, J. F., Guest, J. E., Schaber, G. G., Trask, N. J. and Greeley, R. (1981). Stratigraphy of the Caloris Basin, Mercury. *Icarus* **47**, 184–202.

Macdonald, G. A. (1967). Forms and structures of extrusive basaltic rocks. In *Basalts*, H. H. Hess and A. Poldervaart (eds.), 1–61. John Wiley Interscience, New York.

Macdonald, G. A. (1972). *Volcanoes*. Prentice-Hall, Englewood Cliffs, NJ.

McEwen, A. S. and Soderblom, L. A. (1983). Two classes of volcanic plumes on Io. *Icarus* **55**, 191–217.

McGetchin, T. R. and Head, J. W. (1973). Lunar cinder cones. *Science* **180**, 68–71.

McGill, G. E., Steenstrup, S. J., Barton, C. and Ford, P. G. (1981). Continental rifting and the origin of Beta Regio, Venus. *Geophys. Res. Lett.* **8**, 737–40.

McKee, E. D. (1979). Introduction to a study of global sand seas. In *A study of global sand seas*, E. D. McKee (ed.). US Geol Surv. Prof. Paper 1052, 1–19.

McKinnon, W. B. and Melosh, H. J. (1980). Evolution of planetary lithospheres: evidence from multiringed structures on Ganymede and Callisto. *Icarus* **44**, 454–71.

McLaughlin, D. (1954). Volcanism and aeolian deposition on Mars. *Geol Soc. Am. Bull.* **65**, 715–17.

Malin, M. C. (1976). Observations of intercrater plains on Mercury. *Geophys. Res. Lett.* **3**, 581–4.

Malin, M. C. (1978). Surfaces of Mercury and the Moon: effects of resolution and lighting conditions on the discrimination of volcanic features. *Proc. Lunar Planet. Sci. Conf. 9th*, 3395–409.

Malin, M. C. (1980). Domes on Ganymede. In *Rep. Planet. Geol. Prog.—1980*, NASA TM-82385, 67.

Malin, M. C. and Dzurisin, D. (1977). Landform degradation on Mercury, the Moon, and Mars: evidence from crater depth/diameter relationships. *J. Geophys. Res.* **82**, 376–88.

Malin, M. C. and Dzurisin, D. (1978). Modification of fresh crater landforms: evidence from the Moon and Mercury. *J. Geophys. Res.* **83**, 233–43.

Masursky, H. (1973). An overview of geological results from Mariner 9. *J. Geophys. Res.* **78**, 4009–30.

Masursky, H., Boyce, J. M., Dial, A. L., Schaber, G. G. and Strobell, M. E. (1977). Classification and time of formation of martian channels based on Viking data. *J. Geophys. Res.* **82**, 4016–38.

Masursky, H., Schaber, G. G., Soderblom, L. A. and Strom, R. G. (1979). Preliminary geological mapping of Io. *Nature* **280**, 725–9.

Masursky, H., Eliason, E., Ford, P. G., McGill, G. E., Pettengill, G. H., Schaber, G. G., and Schubert, G. (1980). Pioneer Venus radar results: geology from images and altimetry. *J. Geophys. Res.* **85**, 8232–60.

Matson, D. L., Ransford, G. A. and Johnson, T. V. (1981). Heat flow from Io. *J. Geophys. Res.* **86**, 1664–72.

MCWG (1983). Channels and valleys on Mars. *Geol Soc. Am. Bull.* **94**, 1035–54.

Melosh, H. J. (1977). Global tectonics of a despun planet. *Icarus* **31**, 221–43.

Melosh, H. J. (1982). A simple mechanical model of Valhalla Basin, Callisto. *J. Geophys. Res.* **87**, 1880–90.

Moore, H. J., Boyce, J. M., Schaber, G. G. and Scott, D. H. (1980). *Lunar remote sensing and measurements*. US Geol Surv. Prof. Pap. 1046-B.

Moore, H. J., Plaut, J. J., Schenk, P. M. and Head, J. W. (1992). An unusual volcano on Venus. *J. Geophys. Res.* **97**, 13 479–94.

Moore, J. M. (1984). The tectonic and volcanic history of Dione. *Icarus* **59**, 205–20.

Moore, J. M. and Ahern, J. L. (1982). Tectonic and geological history of Tethys. *Lunar Planet. Sci.* **13**, 538–9.

Moore, J. M. and Ahern, J. L. (1983). The geology of Tethys. *J. Geophys. Res.* **88**, A577–84.

Moore, J. M. and Horner, V. M. (1984). The geomorphologic features on Rhea. *Lunar and Planet. Sci.* **15**, 560–1.

Moore, J. M., Horner, V. M. and Greeley, R. (1985). Geomorphology of Rhea: implications for geologic history and surface processes. *J. Geophys. Res.* **90**, 785–96.

Morabito, L. A., Synnott, S. P., Kupferman, P. N. and Collins, S. A. (1979). Discovery of currently active extraterrestrial volcanism. *Science* **204**, 972.

Morris, E. C. (1982). Aureole deposits of the martian volcano Olympus Mons. *J. Geophys. Res.* **87**, 1164–78.

Morris, E. C. and Underwood, J. R. (1978). Polygonal fractures of the martian plains. In *Rep. Planet. Geol. Prog. 1977–1978*, NASA TM-79729, 97–9.

Morrison, D. (1982a). Introduction to the satellites of Jupiter. In *Satellites of Jupiter*, D. Morrison (ed.), 3–43. University of Arizona Press, Tucson.

Morrison, D. (1982b). *Voyages to Saturn*. NASA SP-451.

Morrison, D. (1983). Outer planets satellites. *US Nat. Rep. 1979–1982*, 151–9.

Morrison, D. and Samz, J. (1980). *Voyage to Jupiter*. NASA SP-439.

Mouginis-Mark, P. (1979). Martian fluidized crater morphology: variations with crater size, latitude, altitude, and target material. *J. Geophys. Res.* **84**, 8011–22.

Mouginis-Mark, P. (1981). Ejecta emplacement and modes of formation of martian fluidized ejecta craters. *Icarus* **45**, 60–76.

Murray, B. C. (1975). The Mariner 10 pictures of Mercury: an overview. *J. Geophys. Res.* **80**, 2342–4.

Murray, B. C. (1983). *The planets—readings from Scientific American*. W. H. Freeman, San Francisco.

Murray, B. C. and Burgess, E. (1977). *Flight to Mercury*. Columbia University Press, New York.

Murray, B. C., Strom, R. G., Trask, N. J. and Gault, D. E. (1975). Surface history of Mercury: implications for terrestrial planets. *J. Geophys. Res.* **80**, 2508–14.

Murray, B. C., Malin, M. C. and Greeley, R. (1981). *Earthlike planets.* W. H. Freeman, San Francisco.

Mutch, T. A., Arvidson, R. E., Head, J. W., Jones, K. L. and Saunders, R. S. (1976). *The geology of Mars.* Princeton University Press, Princeton.

Nelson, R. M., Pieri, D. C., Nash, D. and Baloga, S. M. (1982). Reflection spectrum of liquid sulfur and its implication for Io. In *Rep. Planet. Geol. Prog.—1982,* NASA TM-85127, 12–15.

Ness, N. F., Acuña, M. H., Burlaga, L. F., Connerney, J. E. P., Lepping, R. P. and Neubauer, F. M. (1989). Magnetic fields at Neptune. *Science* **246**, 1473–8.

Neukum, G. and Hiller, K. (1981). Martian ages. *J. Geophys. Res.* **86**, 3097–121.

Neukum, G. and Wise, D. U. (1976). Mars: a standard crater curve and possible new time scale. *Science* **194**, 1381–7.

Newburn, R. (1978). Planetary data. *Proc. Lunar Planet. Sci. Conf. 9th,* end papers.

Newburn, R. and Matson, D. (1978). Satellite data. *Proc. Lunar Planet. Sci. Conf. 9th,* end papers.

Nummedal, D. (1978). The role of liquefaction in channel development on Mars. In *Rep. Planet. Geol. Prog.—1977–1978,* NASA TM-79729, 257–9.

Oberbeck, V. R. and Quaide, W. L. (1968). Genetic implications of lunar regolith thickness variations. *Icarus* **9**, 446–65.

Oberbeck, V. R., Quaide, W. L., Arvidson, R. E. and Aggarwal, H. R. (1977). Comparative studies of lunar, martian, and mercurian craters and plains. *J. Geophys. Res.* **82**, 1681–98.

Orton, G. (1978). Planetary atmospheres. *Proc. Lunar Planet. Sci. Conf. 9th,* end papers.

Owen, T. (1983). Titan. In *The planets,* B. Murray (ed.), 84–93. W.H. Freeman, San Francisco.

Parmentier, E. M. and Head, J. W. (1981). Viscous relaxation of impact craters on icy planetary surfaces: determination of viscosity variation with depth. *Icarus* **47**, 100–11.

Parmentier, E. M., Squyres, S. W., Head, J. W. and Allison, M. L. (1982). The tectonics of Ganymede. *Nature* **295**, 290–3.

Passey, Q. R. (1983). Viscosity of the lithosphere of Enceladus. *Icarus* **53**, 105–20.

Passey, Q. R. and Shoemaker, E. M. (1982). Craters and basins on Ganymede and Callisto: morphological indicators of crustal evolution. In *Satellites of Jupiter,* D. Morrison (ed.), 379–434. University of Arizona Press, Tucson.

Pavri, B., Head, J. W., Klose, K. B. and Wilson, L. (1992).

Steep-sided domes on Venus: characteristics, geologic setting, and eruption conditions from Magellan data. *J. Geophys. Res.* **97**, 13,445–78.

Peale, S. J., Cassen, P. and Reynolds, R. T. (1979). Melting of Io by tidal dissipation. *Science* **203**, 892–4.

Pechmann, J. C. (1980). The origin of polygonal troughs on the northern plains of Mars. *Icarus* **42**, 185–210.

Peterson, J. E. (1978). Antipodal effects of major basin-forming impacts on Mars. *Lunar Planet. Sci.* **9**, 885–6.

Pettengill, G. H., Eliason, E., Ford, P. G., Loriot, G. B., Masursky, H. and McGill, G. E. (1980). Pioneer Venus radar results: altimetry and surface properties. *J. Geophys. Res.* **85**, 8261–70.

Pettengill, G. H., Ford, P. G., Johnson, W. T. K., Raney, R. K., and Soderblom, L. A. (1991). Magellan: radar performance and data products. *Science* **252**, 260–5.

Phillips, R. J. and Ivins, E. R. (1979). Geophysical observations pertaining to solid-state convection in the terrestrial planets. *Phys. Earth Planet. Interiors* **19**, 107–48.

Phillips, R. J. and Lambeck, K. (1980). Gravity fields of the terrestrial planets: long-wavelength anomalies and tectonics. *Rev. Geophys. Space Phys.* **18**, 27–76.

Phillips, R. J., Arvidson, R. E., Boyce, J. M., Campbell, D. B., Guest, J. E., Schaber, G. G. and Soderblom, L. A. (1991). Impact crater on Venus: Initial analysis from Magellan. *Science* **252**, 288–97.

Phillips, R. J., Raubertas, R. F., Arvidson, R. E., Sarkar, I. C., Herrick, R. R., Izenberg, N. and Grimm, R. E. (1993). Impact craters and Venus resurfacing history. *J. Geophys. Res.* (to appear).

Pieri, D. C. (1979), *Geomorphology of martian valleys.* Ph.D. dissertation, Cornell University.

Pieri, D. C. (1980). Lineament and polygon patterns on Europa. *Nature* **289**, 17–21.

Pieri, D. (1983). The ancient rivers of Mars. *The Planet. Rep.* **3**, 4–7.

Pieri, D. C., Baloga, S. M. and Nelson, R. M. (1982). Colors of lava flows at Ra Patera, Io. In *Rep. Planet. Geol. Prog.—1982,* NASA TM-85127, 16–19.

Pieri, D. C., Baloga, S. M., Nelson, R. M. and Sagan, C. (1984). Sulfur flows of Ra Patera, Io. *Icarns* **60**, 685–700.

Pike, R. J. (1974). Depth/diameter relations of fresh lunar craters: revision from spacecraft data. *Geophys. Res. Lett.* **1**(7), 291–4.

Pike, R. J. (1976). *Geologic map of the Rima Hyginus region of the Moon.* US Geol Surv. Misc. Inv. Map I-945.

Pike, R. J. (1980). *Apollo 15–17 orbital investigations—geometric interpretation of lunar craters.* US Geol Surv. Prof. Paper 1046C.

Pilcher, C. B., Ridgway, S. T. and McCord, T. B. (1972). Galilean satellites: identification of water frost. *Science* **178**, 1087–9.

Plaut, J. J. and Arvidson, R. E. (1993). Comparison of

goldstone and Magellan radar data in the equatorial plains of Venus. *J. Geophys. Res.* (to appear).

Plescia, J. B. (1983). The geology of Dione. *Icarus* **56**, 255–77.

Plescia, J. B. and Boyce, J. M. (1982). Crater densities and geological histories of Rhea, Dione, Mimas and Tethys. *Nature* **295**, 285–90.

Plescia, J. B. and Boyce, J. M. (1983). Crater numbers and geological histories of Iapetus, Enceladus, Tethys and Hyperion. *Nature* **301**, 666–70.

Plescia, J. B. and Saunders, R. S. (1979). The chronology of martian volcanoes. *Proc. Lunar Planet. Sci. Conf. 10th,* 2841–59.

Poirier, J. P. (1982). Rheology of ices: a key to the tectonics of the ice moons of Jupiter and Saturn. *Nature* **299**, 683–8.

Pollack, J. B. and Sagan, C. (1965). The microwave phase effect of Venus. *Icarus* **4**, 62–103.

Pollack, J. B., Burns, J. A. and Tauber, M. E. (1979). Gas drag in primordial circumplanetary envelopes: a mechanism for satellite capture. *Icarus* **37**, 587–611.

Ransford, G. A., Finnerty, A. A. and Collerson, K. D. (1980). Europa's petrological thermal history. *Nature* **289**, 21–4.

Rittmann, A. (1962). *Volcanoes and their activity* (translated by E. A. Vincent from the German 2nd edn). Interscience, New York.

Roberts, K. M., Guest, J. E., Head, J. W. and Lancaster, M. G. (1992). Mylitta Fluctus, Venus: rift-related, centralized volcanism and the emplacement of large-volume flow units. *J. Geophys. Res.* **97**, 15 991–16 016.

Roddy, D. J. (1977). Large-scale impact and explosion craters: comparisons of morphological and structural analogs. In *Impact and explosion cratering: planetary and terrestrial implications.* D. J. Roddy, R. O. Pipen and R. B. Merrill (eds.), 185–246. Pergamon Press, New York.

Roddy, D. J., Pepin, R. O. and Merrill, R. B. (eds) (1977). *Impact and explosion cratering: planetary and terrestrial implications.* Pergamon Press, New York.

Rossbacher, L. A. and Judson, S. (1981). Ground ice on Mars: inventory, distribution, and resulting landforms. *Icarus* **45**, 39–59.

Ryder, G. and Spudis, P. D. (1980). Volcanic rocks in the lunar highlands. *Proc. Conf. Lunar Highlands Crust,* J. J. Papike and R. B. Merrill (eds.), 353–75. Pergamon Press, New York.

Ryder, G. and Taylor, G. J. (1976). Did mare-type volcanism commence early in lunar history? *Proc. Lunar Planet. Sci. Conf. 7th,* 1741–55.

Sagan, C. (1961). The planet Venus. *Science* **133**, 849–58.

Sagan, C. (1979). Sulfur flows on Io. *Nature* **280**, 750–3.

Sagan, C. and Dermott, S. F. (1982). The tide in the seas of Titan. *Nature* **300**, 731–3.

Sagan, C., Veverka, J., Fox, P., Dubisch, R., Lederberg, J., Levinthal, E., Quam, L., Tucker, R., Pollack, J. B. and Smith, B. A. (1972). Variable features on Mars: preliminary Mariner 9 television results. *Icarus* **17**, 346–72.

Sagan, C., Veverka, J., Fox, P., Dubisch, R., French, R., Gierasch, P., Quam, L., Lederberg, J., Levinthal, E., Tucker, R., Eross, B. and Pollack, J. B. (1973). Variable features on Mars, 2, Mariner 9 global results. *J. Geophys. Res.* **78**, 4163–96.

Sandwell, D. T. and Schuberg, G. (1992). Evidence for retrograde lithospheric subduction on Venus. *Science* **257**, 766–70.

Saunders, R. S. and Pettengill, G. H. (1991). Magellan: Mission summary. Science **252**, 247–9.

Saunders, R. S., Arvidson, R. E., Head, J. W., Schaber, G. G., Stofan, E. R. and Solomon, S. C. (1991). An overview of Venus geology. *Science* **252**, 249–52.

Schaber, G. G. (1973). Lava flows in Mare Imbrium: geologic evaluation from Apollo orbital photography. *Proc. Lunar Sci. Conf. 4th,* 73–92.

Schaber, G. G. (1980). The surface of Io: geologic units, morphology and tectonics. *Icarus* **43**, 302–33.

Schaber, G. G. (1982*a*). Syrtis Major: a low-relief volcanic shield. *J. Geophys. Res.* **87**, 9852–66.

Schaber, G. G. (1982*b*). The geology of Io. In *Satellites of Jupiter.* D. Morrison (ed.), 556–97. University of Arizona Press, Tucson.

Schaber, G. G., Boyce, J. M. and Trask, N. J. (1977). Moon–Mercury: large impact structures, isostasy and average crustal viscosity. *Phys. Earth Planet. Int.* **15**, 189–201.

Schaber, G. G., Strom, R. G., Moore, H. J., Soderblom, L. A., Kirk, R. L., Chadwick, D. J., Dawson, D. D., Gaddis, L. R., Boyce, J. M. and Russel, J. (1992). Geology and distribution of impact craters on Venus: What are they telling us? *J. Geophys. Res.* **97**, 13 257–301.

Schonfeld, E. (1977). Martian volcanism. *Lunar Sci.* **8**, 843–5.

Schultz, P. H. (1976*a*). Floor-fractured craters. *Moon* **15**, 241–73.

Schultz, P. H. (1976*b*). Introduction to planetary surfaces. In *NASA short course in planetary geology,* P. H. Schultz and R. Greeley (eds.), 1–14.

Schultz, P. H. (1976*c*). *Moon morphology.* University of Texas Press, Austin.

Schultz, P. H. (1977). Endogenic modification of impact craters on Mercury. *Phys. Earth Planet. Interiors* **15**, 202–19.

Schultz, P. H. (1981). Impact cratering on Venus. *Int. Conf. Venus Environment,* 6.

Schultz, P. H. and Gault, D. E. (1975). Seismic effects from major basin formation on the Moon and Mercury. *Moon* **12**, 159–77.

Schultz, P. H. and Gault, D. E. (1979). Atmospheric effects

on martian ejecta emplacement. *J. Geophys. Res.* **84**, 7669–87.

Schultz, P. H. and Lutz-Garihan, A. B. (1982). Grazing impacts on Mars: a record of lost satellites. *J. Geophys. Res.* **87**, A84–A96.

Schultz, P. H. and Spudis, P. D. (1979). Evidence for ancient mare volcanism. *Proc. Lunar Planet. Sci. Conf. 10th*, 2899–918.

Schultz, P. H. and Spudis, P. D. (1983). Beginning and end of lunar mare volcanism. *Nature* **302**, 233–6.

Schultz, P. H., Schultz, R. A. and Rogers, J. (1982). The structure and evolution of ancient impact basins on Mars. *J. Geophys. Res.* **87**, 9803–20.

Scott, D. H. (1977). Moon–Mercury: relative preservation states of secondary craters. *Phys. Earth Planet. Int.* **15**, 173–8.

Scott, D. H. (1982). Volcanoes and volcanic provinces: Martian western hemisphere. *J. Geophys. Res.* **87**, 9839–51.

Scott, D. H. and Carr, M. H. (1978). *Geologic map of Mars.* US Geol Survey, Misc. Inv. Map I-1083.

Scott, D. H. and Tanaka, K. L. (1981). Mars: paleostratigraphic restoration of buried surfaces in Tharsis Montes. *Icarus* **45**, 304–19.

Scott, D. H., McCauley, J. F. and West, M. N. (1978). *Geologic map of the west side of the Moon.* US Geol Surv. Misc. Geol. Inv. Map I-1034.

Sharp, R. P. (1973). Mars: south polar pits and etched terrains. *J. Geophys. Res.* **78**, 4222–30.

Sharp, R. P. (1980). Geomorphological processes on terrestrial planetary surfaces. *Ann. Rev. Earth Planet. Sci.* **8**, 231–61.

Sharp, R. P. and Malin, M. C. (1975). Channels on Mars. *Geol Soc. Am. Bull.* **86**, 593–609.

Sharpe, C. F. S. (1968). *Landslides and related phenomena.* Cooper Square, New York.

Shoemaker, E. M. (1963). Impact mechanics at Meteor Crater, Arizona. In *The Solar System, the moon, meteorites and comets,* vol. 4, B. M. Middlehurst and G. P. Kuiper (eds), 301–36. University of Chicago Press, Chicago.

Shoemaker, E. M. and Hackman, R. J. (1962). Stratigraphic basis for a lunar time scale. In *The Moon—International Astronomical Union Symposium 14,* A. Kopal and Z. K. Mikhailov (eds), 289–300. Academic Press, New York.

Shoemaker, E. M. and Wolfe, R. F. (1982). Cratering time scales for the Galilean satellites. In *Satellites of Jupiter,* D. Morrison (ed.), 277–339. University of Arizona Press, Tucson.

Shoemaker, E. M., Lucchitta, B. K., Plescia, J. B., Squyres, S. W. and Wilhelms, D. E. (1982). The geology of Ganymede. In *Satellites of Jupiter,* D. Morrison (ed.), 435–520. University of Arizona Press, Tucson.

Sill, G. T. and Clark, R. N. (1982). Composition of the surface of the Galilean satellites. In *Satellites of Jupiter,* D. Morrison (ed.), 174–212. University of Arizona Press, Tucson.

Silver, L. T. and Schultz, P. H. (eds) (1982). *Geological implications of impacts of large asteroids and comets on the Earth.* Geol Soc. Am. Spec. Paper 190.

Sjogren, W. L. (1979). Mars gravity: high resolution results from Viking orbiter 2. *Science,* **203**, 1006–9.

Sjogren, W. L., Lorell, J., Wong, L. and Downs, W. (1975). Mars gravity field based on a short-arc technique. *J. Geophys. Res.* **80**, 2899–908.

Sleep, N. H. and Phillips, R. J. (1979). An isostatic model for the Tharsis province, Mars, *Geophys. Res. Lett.* **6**, 803–6.

Smith, B. A., Soderblom, L. A., Beebe, R., Boyce, J., Briggs, G., Carr, M., Collins, S. A., Cook II, A. F., Danielson, G. E., Davies, M. E., Hunt, G. E., Ingersoll, A., Johnson, T. V., Masursky, H., McCauley, J., Morrison, D., Owen, T., Sagan, C., Shoemaker, E. M., Strom, R., Suomi, V. E. and Veverka, J. (1979*a*). The Galilean satellites and Jupiter: Voyager 2 imaging science results. *Science* **206**, 927–50.

Smith, B. A., Soderblom, L., Johnson, T. V., Ingersoll, A. P., Collins, S. A., Shoemaker, E. M., Hunt, G. E., Masursky, H., Carr, M. H., Davies, M. E., Cook II, A. F., Boyce, J., Danielson, G. E., Owen, T., Sagan, C., Beebe, R. F., Veverka, J., Strom, R. G., McCauley, J. F., Morrison, D., Briggs, G. A. and Suomi, V. E. (1979*b*). The Jupiter system through the eyes of Voyager 1. *Science* **204**, 951–72.

Smith, B. A., Soderblom, L., Beebe, R., Boyce, J., Briggs, G., Bunker, A., Collins, S. A., Hansen, C. J., Johnson, T. V., Mitchell, J. L., Terrile, R. J., Carr, M., Cook II, A. F., Cuzzi, J., Pollack, J. B., Danielson, G. E., Ingersoll, A., Davies, M. E., Hunt, G. E., Masursky, H., Shoemaker, E. and Morrison, D. (1981). Encounter with Saturn: Voyager 1 imaging science results. *Science* **212**, 163–91.

Smith, B. A., Soderblom, L., Batson, R., Bridges, P., Inge, J., Masursky, H., Shoemaker, E., Beebe, R., Boyce, J., Briggs, G., Bunker, A., Collins, S. A., Hansen, C. J., Johnson, T. V., Mitchell, J. L., Terrile, R. J., Cook II, A. F., Cuzzi, J., Pollack, J. B., Danielson, G. E., Ingersoll, A. P., Davies, M. E. and Hunt, G. E. (1982). A new look at the Saturn system: the Voyager 2 images. *Science* **215**, 504–37.

Smith, B. A., Soderblom, L. A., Beebe, R., Bliss, D., Boyce, J. M., Briggs, G. A., Brown, R. H., Collins, S. A., Cook II, A. F., Croft, S. K., Cuzzi, J. N., Danielson, G. E., Davies, M. E., Dowling, T. E., Godfrey, D., Hansen, C. J., Harris, C., Hunt, G. E., Ingersoll, A. P., Johnson, T. V., Krauss, R. J., Masursky, H., Morrison, D., Owen, T., Plescia, J. B., Pollack, J. B., Porco, C. P., Rages, K., Sagan, C., Shoemaker, E. M.,

Sromovsky, L. A., Stroker, C., Strom, R. G., Suomi, V. E., Synnott, S. P., Terrile, R. J., Thomas, P., Thompson, W. R. and Veverka, J. (1986). Voyager 2 in the Uranian system: imaging science results. *Science* **233**, 43–64.

Smith, B. A. and the Voyager imaging team (1989). Voyager 2 at Neptune: Imaging science results. *Science* **246**, 1422–49.

Smith, E. I. (1976). Comparison of the crater morphology-size relationship for Mars, Moon, and Mercury. *Icarus* **28**, 543–50.

Smith, R. W. (1982). The impact on astronomy of the discovery of Uranus. In *Uranus and the outer planets*, G. Hunt (ed.) 81–9. Cambridge University Press, Cambridge.

Soderblom, L. A. (1980). The Galilean moons of Jupiter. *Scientific Am.* **242**, 88–100.

Soderblom, L. and Johnson, T. V. (1983). The moons of Saturn. In *The planets*, B. Murray (ed.), 95–107. W. H. Freeman, San Francisco.

Soderblom, L. A. and Wenner, D. B. (1978). Possible fossil H_2O liquid–ice interfaces in the martian crust. *Icarus* **34**, 622–37.

Soderblom, L. A., West, R. A., Herman, B. M., Kreidler, T. J. and Condit, C. D. (1974). Martian planetwide crater distributions: implications for geologic history and surface processes. *Icarus* **22**, 239–63.

Soderblom, L. A., Kieffer, S. W., Becker, T. L., Brown, R. H., Cook II, A. F., Hansen, C. J., Johnson, T. V., Kirk, R. L. and Shoemaker, E. M. (1990). Triton's geyser-like plumes: discovery and basic characterization. *Science* **250**, 41–415.

Solomon, S. C. (1976). Some aspects of core formation in Mercury. *Icarus* **28**, 509–21.

Solomon, S. C. (1981). The geophysics of Mars: whence the Tharsis plateau? *Nature* **294**, 304–5.

Solomon, S. C. and Head, J. W. (1982*a*). Evolution of the Tharsis province of Mars: the importance of heterogeneous lithospheric thickness and volcanic construction. *J. Geophys. Res.* **87**, 9755–74.

Solomon, S. C. and Heads, J. W. (1982*b*). Mechanisms for lithospheric heat transport on Venus: implications for tectonic style and volcanism. *J. Geophys. Res.* **87**, 9236–46.

Solomon, S. C., Head, J. W., Kaula, W. M., McKenzie, D., Parsons, B., Phillips, R. J., Schubert, G. and Talwani, M. (1991). Venus tectonics: Initial analysis from Magellan. *Science* **252**, 297–312.

Solomon, S. C., Smrekar, S. E., Bindschadler, D. L., Grimm, R. E., Kaula, W. M., McGill, G. E., Phillips, R. J., Saunders, R. S., Schubert, G., Squyres, S. W. and Stofan, E. R. (1992). Venus tectonics: an overview of Magellan observation. *J. Geophys. Res.* **97**, 13 199–256.

Spudis, P. (1978). Composition and origin of the Apennine Bench Formation. *Proc. Lunar Planet. Sci. Conf. 9th,* 3379–94.

Spudis, P. D. (1982). The geology of lunar multi-ring basins. Unpublished Ph.D. dissertation, Arizona State University.

Spudis, P. and Greeley, R. (1976). *Surficial geology of Mars: a study in support of a penetrator mission to Mars.* NASA TM X-73184.

Spudis, P. D. and Strobell, M. E. (1984). New identification of ancient multi-ring basins on Mercury and implications for geologic evolution. *Lunar Planet. Sci.* **15**, 814–15.

Spurr, J. E. (1944). *Geology applied to selenology: I. the Imbrium plain region of the Moon.* Science Press, Lancaster.

Spurr, J. E. (1945). *Geology applied to selenology: II. the features of the Moon.* Science Press, Lancaster.

Spurr, J. E. (1948). *Geology applied to selenology: III. lunar catastrophic history.* Rumford Press, Concord.

Spurr, J. E. (1949). *Geology applied to selenology: IV, the shrunken Moon.* Rumford Press, Concord.

Squyres, S. W. (1979). The distribution of lobate debris aprons and similar flows on Mars. *J. Geophys. Res.* **84**, 8087–96.

Squyres, S. W. (1980). Volume changes in Ganymede and Callisto and the origin of grooved terrain. *Geophys. Res. Lett.* **7**, 593–6.

Squyres, S. W. (1989). Urey prize lecture: Water on Mars. *Icarus* **79**, 229–88.

Squyres, S. W., Reynolds, R. T., Cassen, P. M. and Peale, S. J. (1983). The evolution of Enceladus. *Icarus* **53**, 319–31.

Squyres, S. W., Janes, D. M., Baer, G., Bindschadler, D. L., Schubert, G., Sharpton, V. L. and Stofan, E. R. (1992). The morphology and evolution of coronae on Venus. *J. Geophys. Res.* **97**, 13 611–34.

SSEC (Solar System Exploration Committee) (1983). *Planetary exploration through the year 2000: a core program.* Washington, DC: NASA.

Stephens, S. K., Solomon, S. C. and Head, J. W. (1983). On the age of Venus highland topography: constraints from the viscous relaxation of relief. *Lunar Planet. Sci.* **14**, 747–8.

Stevenson, D. J. (1982). Volcanism and igneous processes in small icy satellites. *Nature* **298**, 142–4.

Stofan, E. R., Sharpton, V. L., Schubert, G., Baer, G., Bindschadler, D. L., Janes, D. M. and Squyres, S. W. (1992). Global distribution and characteristics of coronae and related features on Venus: Implications for origin and relation to mantle processes. *J. Geophys. Res.* **97**, 13 347–78.

Stone, E. C. and Miner, E. D. (1989). The Voyager 2 encounter with the Neptunian System. *Science* **246**, 1417–21.

Strain, P. L. and El-Baz, F. (1980). The geology and morphology of Ina. *Proc. Lunar Planet. Sci. Conf. 11th,* 2437–46.

Strom, R. G. (1979). Mercury: a post-Mariner 10 assessment. *Space Sci. Rev.* **24**, 3–70.

Strom, R. G. and Schneider, N. M. (1982). Volcanic eruption plumes on Io. In *Satellites of Jupiter.* D. Morrison (ed.), 598–633. University of Arizona Press, Tucson.

Strom, R. G., Murray, B. C., Belton, M. J. S., Danielson, G. E., Davies, M. E., Gault, D. E., Hapke, B., O'Leary, B., Trask, N., Guest, J. E., Anderson, J. and Klaasen, K. (1975a). Preliminary imaging results from the second Mercury encounter. *J. Geophys. Res.* **80**, 2345–56.

Strom, R. G., Trask, N. J. and Guest, J. E. (1975b). Tectonism and volcanism on Mercury. *J. Geophys. Res.* **80**, 2478–507.

Strom, R. G., Terrile, R. J., Masursky, H. and Hansen, C. (1979). Volcanic eruption plumes on Io. *Nature* **280**, 733–6.

Strom, R. G., Croft, S. K. and Boyce, J. M. (1990). The impact cratering record on Triton. *Science* **250**, 437–9.

Stuart-Alexander, D. E. and Howard, K. A. (1970). Lunar maria and circular basins—a review. *Icarus* **12**, 440–56.

Surkov, Y. A. (1983). Studies of Venus rocks by Veneras 8, 9 and 10. In *Venus,* D. M. Hunten, L. Colin, T. M. Donahue and V. I. Moroz (eds.), 154–8. University of Arizona Press, Tucson.

Tauber, M. E. and Kirk, D. B. (1976). Impact craters on Venus. *Icarus* **28**, 351–7.

Tauber, M. E., Kirk, D. B. and Gault, D. E. (1978). An analytic study of impact ejecta trajectories in the atmospheres of Venus, Mars, and Earth. *Icarus* **33**, 529–36.

Taylor, S. R. (1982). *Planetary science: a lunar perspective.* Lunar Planetary Institute, Houston.

Terrile, R. J., Johnson, T. V., Soderblom, L. A. and Strom, R. G. (1981). Variable features on Io. In *Rep. Planet. Geol. Prog.—1981,* NASA TM-84211, 29–31.

Theilig, E. and Greeley, R. (1979). Plains and channels in the Lunae Planum—Chryse Planitia region of Mars. *J. Geophys. Res.* **84**, 7994–8010.

Thomas, P. (1980). Plain formation on Mercury: tectonic implications. *Moon Planets* **22**, 261–8.

Thomas, P. and Veverka, J. (1979). Seasonal and secular variation of wind streaks on Mars: an analysis of Mariner 9 and Viking data. *J. Geophys. Res.* **84**, 8131–46.

Thomas, P. and Veverka, J. (1982). Amalthea. In *Satellites of Jupiter,* D. Morrison (ed.), 147–73. University of Arizona Press, Tucson.

Thomas, P., Veverka, J. and Duxbury, T. (1978). Origin of the grooves on Phobos. *Nature* **273**, 282–4.

Thomas, P. G., Masson, P. and Fleitout, L. (1982). Global volcanism and tectonism on Mercury: comparison with the Moon. *Earth Planet. Sci. Lett.* **58**, 95–103.

Thompson, W. R. and Sagan, C. (1990). Color and chemistry on Triton. *Science* **250**, 415–18.

Trask, N. J. and Guest, J. E. (1975). Preliminary geologic terrain map of Mercury. *J. Geophys. Res.* **80**, 2461–77.

Tsoar, H., Greeley, R. and Peterfreund, A. R. (1979). Mars: the north polar sand sea and related wind patterns. *J. Geophys. Res.* **84**, 8167–82.

Twidale, C. R. (1976). *Analysis of landforms.* John Wiley, Brisbane.

Tyler, G. L., Eshleman, V. R., Anderson, J. D., Levy, G. S., Lindal, G. F., Wood, G. E. and Croft, T. A. (1981). Radio science investigations of the Saturn system with Voyager 1: preliminary results. *Science* **212**, 201–6.

Urey, H. C. (1952). *The planets: their origin and development.* Yale University Press, New Haven.

Varnes, D. J. (1978). Slope movement types and processes. In *Landslides analysis and control,* R. L. Schuster and R. J. Krisek (eds), 11–33. National Academy of Sciences, Washington, DC.

Veverka, J. and Thomas, P. (1979). Phobos and Deimos: a preview of what asteroids are like? In *Asteroids,* T. Gehrels (ed.), 628–51. University of Arizona Press, Tucson.

Walker, G. P. L. (1973). Lengths of lava flows. *Phil Trans R. Soc. Lond.* A **274**, 107–18.

Ward, A. W. (1979). Yardangs on Mars: evidence of recent wind erosion. *J. Geophys. Res.* **84**, 8147–66.

Warner, J. L. (1983). Sedimentary processes and crustal cycling on Venus. *J. Geophys. Res.* **88**, suppl. A495–A500.

Wentworth, C. K. and Macdonald, G. A. (1953). Structures and forms of basaltic rocks in Hawaii. *US Geol Surv. Bull.* **994.**

Whitaker, E. A. (1972). An unusual mare feature. In *Apollo 15 Prelim. Sci. Rep.* NASA SP-289, 25–84 through 25–85.

Whitford-Stark, J. L. (1982). Factors influencing the morphology of volcanic landforms: an Earth–Moon comparison. *Earth Sci. Rev.* **18**, 109–68.

Wilhelms, D. E. (1976). Mercurian volcanism questioned. *Icarus* **28**, 551–8.

Wilhelms, D. E. (1980). *Stratigraphy of part of the lunar nearside.* US Geol Surv. Prof. Paper 1046-A.

Wilhelms, D. E. and Davis, D. E. (1971). Two former faces of the Moon. *Icarus* **15**, 368–72.

Wilhelms, D. E. and McCauley, J. F. (1971). *Geologic map of the nearside of the Moon.* US Geol Surv. Misc. Geol. Inv. Map I-703.

Wilshire, H. G., Offield, T. W., Howard, K. A. and Cummings, D. (1972). *Geology of the Sierra Madera cryptoexplosion structure, Pecos County, Texas.* US Geol Surv. Prof. Paper 599-H.

Wilson, L. and Head, J. W. (1981). Ascent and eruption of basaltic magma on the Earth and Moon. *J. Geophys. Res.* **86**, 2971–3001.

Wise, D. U., Golombek, M. P. and McGill, G. E. (1979). Tectonic evolution of Mars. *J. Geophys. Res.* **84**, 7934–9.

Wood, C. A. and Head, J. W. (1976). Comparison of impact basins on Mercury, Mars, and the Moon. *Proc. Lunar Sci. Conf. 7th*, 3629–51.

Wood, C. A., Head, J. W. and Cintala, M. J. (1978). Interior morphology of fresh martian craters: the effects of target characteristics. *Proc. Lunar Planet. Sci. Conf. 9th*, 3691–709.

Woronow, A., Strom, R. G. and Gurnis, M. (1982). Interpreting the cratering record: Mercury to Ganymede and Callisto. In *Satellites of Jupiter*, D. Morrison (ed.), 237–76. University of Arizona Press, Tucson.

Young, R. A., Brennan, W. J., Wolfe, R. W. and Nichols, D. J. (1973). Volcanism in the lunar maria. In *Apollo 17 Prelim. Sci. Rep.* NASA SP-330, 31-1 through 31-11.

Index*

*Numbers in *italics* refer to text figures.